MW01641393

NAVAL ENGINEER'S GUIDE

NAVAL ENGINEER'S GUIDE

by

Commander James V. Jolliff, U. S. Navy

and

Commander Hollis E. Robertson, U. S. Navy

NAVAL INSTITUTE PRESS

ANNAPOLIS, MARYLAND

Library of Congress Catalog Card No. 78-188009
ISBN 0-87021-415-2

PRINTED IN THE UNITED STATES OF AMERICA

Preface

In January 1911 the first edition of *Watch Officer's Manual,* written by Ensign Charles Emerson Hovey, U. S. Navy, was published. For three generations this definitive reference work—known since 1930 as *Watch Officer's Guide*—has provided valuable guidance to thousands of officers aspiring to qualify as officer of the deck of a naval vessel. Such wide acceptance proves there was a clear need for a manual such as Ensign Hovey produced.

Naval Engineer's Guide has been written to meet an equally clear need in the field of shipboard engineering. Its objective is to provide a compact volume of practical knowledge for both the engineering novice and the old hand. Some attention is paid to engineering watch standing, but the book emphasizes lessons learned from experience and helpful hints concerning routine maintenance and repair, overhauls, refresher training, damage and casualty control, administration, and a multitude of other subjects which daily confront the shipboard engineer.

Whether you are a qualified engineering officer of the watch on an attack aircraft carrier, a new ensign reporting aboard your first destroyer, or a seasoned first class machinist's mate on a coastal minesweeper, *Naval Engineer's Guide* has something of value to offer you.

Acknowledgments

This Guide was prepared under the direction of Mr. G. D. Dunlap, President of Weems & Plath, Inc., Annapolis, Maryland, and the authors are grateful to him for his patience, support, and guidance.

Special thanks are extended to CWO3 Robert F. Carlton, U. S. Navy, for his contribution on electrical systems.

Any opinions and assertions contained in this book are those of the authors and are not to be construed as reflecting the views of the U. S. Navy or the naval service as a whole.

TABLE OF CONTENTS

Chapter

NAVAL ENGINEER'S GUIDE

CHAPTER 1

Introduction

This book is not an engineering textbook, nor is it a replacement for *Naval Ships Technical Manual* (hereinafter referred to as NavShips *Technical Manual*), NavPers training manuals, instruction booklets, operation manuals, or type commanders' instructions. Its purpose is to guide engineering officers and top watch-standing enlisted personnel in the performance of their duties. Hopefully, it will provide the answers to many practical problems encountered by the shipboard engineer who is short on experience or who may have forgotten some of the tricks of the trade during tours ashore or in other shipboard departments. The text deals with everyday, nuts-and-bolts evolutions, and is intended to be helpful particularly to young, inexperienced officers assigned to duty in the engineering department for the first time.

FIGHTING EFFICIENCY

The fighting efficiency of every ship depends on her ability to make the most hits in the shortest possible time with the least expenditure of ammunition, and to make the required number of revolutions with a minimum expenditure of fuel, lubricating oil,

> and water without a breakdown. To make this possible, in time of action, each man, whatever his rating, must be constantly alert and at all times doing his duty, no matter how slight the duty is nor where it is performed. From knowledge, power; from practice, expert performance; and from these, victory.

That quotation is taken from *Handy Book for the Enlisted Men in the Engineering Department, United States Navy,* written by Passed Midshipman Bruce R. Ware, U. S. Navy, and published by the U. S. Naval Institute in 1908. The experiences encountered in ships of the U. S. Navy over the past half-century dramatically confirm the fundamental truth of his statement.

SHIPBOARD ENGINEERING TODAY

The U. S. Navy has progressed considerably since 1908 in the field of marine propulsion. On the horizon we see gas turbine, combined gas turbine and diesel (CODAG), combined steam and gas turbine (COSAG), nuclear, and water-jet plants as modes of propulsion for the future. Furthermore, in the field of merchant ships, great progress has been made in the design of steam power plants. Power plants in the range of 60,000 shaft horsepower (SHP) per shaft and specific fuel rate (SFR), of .46lbs/shp-hr are common. Thus, it is apparent that, in future naval ship designs, steam power plants will be able to compete quite favorably with more exotic combinations, provided the ships are not to be limited by size and weight. The most promising areas for their use would be noncombatant auxiliaries and tenders. The following table, based on a constant displacement of 3,900 tons, a minimum endurance of 5,000 miles at 20 knots, and a variable maximum speed, has been drawn for the purpose of comparison. All the plants listed are currently installed or are under study for application to various destroyer-type ships. Nuclear power has been a mode of submarine propulsion for years, and a small minority of surface ships are propelled in this manner.

Plant	SHP	Top Speed in knots	Range at 20 knots in miles	Relative ship cost	Note No.
Pressure-fired	39,800	30	5,000	1.0	—
Pressure-fired with regeneration	39,800	30	5,050	1.0+	—
Conventional boiler	20,000	26.5	5,000	0.9+	*1*
Diesel	16,000	25	6,000	0.8+	*2*
Gas turbine	33,000	30	5,000	1.0+	*3*
Nuclear	12,000		50,000+	1.5+	*4*

Note 1. Because of the weight of the conventional boiler plant, the power would have to be reduced to hold the ship displacement under 3,900 tons. Some fuel might be traded for machinery weight to permit more power to be installed, but then the ship would not be able to make the required cruising range.

Note 2. Because of weight and space considerations, a diesel plant of 16,000 SHP is about the largest that could be fitted into this selected destroyer-type ship. Thus, for the hull form selected, the maximum ship speed would be about 25 knots. The ship's range would increase because of the higher efficiency of the diesel plant and because the ship would have more volume available *low* in the hull to carry additional fuel.

Note 3. This plant is based on the use of a long-life, efficient gas turbine for cruising speed. It would employ aircraft-type gas turbines for boosting the plant to full power. Although this plant has a high fuel rate, the light weight of the gas turbines would permit carrying a greater fuel load, and the end result would be an increase in endurance.

Note 4. The largest nuclear plant that could be fitted into this selected destroyer-type hull would be about 12,000 SHP. This would result in a reduction of top speed, but a tremendous increase in endurance.

It should be noted that the development of such a table induces some artificial generalities. As an example, if hull displacement were not fixed at 3,900 tons, the conventional boiler plant could be placed in a slightly larger ship and would have the same range and speed as the pressure-fired plant, with only a slight increase in first costs, since hull structure is relatively cheap.

The above discussion does not, of course, cover all the methods that might be used to convert thermal energy to mechanical energy in naval applications. Thermoelectric, magnetohydrodynamic, and thermionic processes are a few of the more exotic systems now under active study by the Navy for application to surface ships and submarines.

Marine power plants are unquestionably headed in the direction of greater reliability, improved maintainability, and increased automation—at least to the extent that automation will lower manning levels and reduce casualties related to operator error while maintaining sufficient personnel to counter battle damage. The question of where marine power plants are headed is inseparably related to the question: "Where are ships headed?" Today the great concern seems to be with increasing the speed of ships; and the power plant, by reason either of the space it occupies or what it costs to run, could very well be the factor that will set maximum speed.

Interesting or unusual as the power plants of the future may be, the fact has to be faced that, at this writing, the majority of the ships in our Navy are over twenty years old, and only a small percentage of them are less than ten years old. In today's surface fleet, the new and the old frequently exist together in a compromise of necessity. It appears less than wise to invest undue confidence in a highly sophisticated electronics system which derives its power solely from overaged turbo-generators that are being supplied with steam from boilers constructed in 1944. New topside weapons, communications, and navigational systems are mounted on old hulls powered by overaged propulsion systems—an unhappy marriage brought about by the restraints of economics and the need to keep our ships afloat to maintain our defense posture.

It would be an understatement to say that the cumulative effects of age are beginning to have an increasingly deleterious effect on the majority of ships now in commission. Hence, it is imperative that engineers do their utmost to become exceptionally proficient in their duties in order to ward off any headlong plunge into block power-plant obsolescence. For that reason, this book deals with the conventional steam power plant currently installed in the majority of this nation's warships and major auxiliary types. Regardless of operating pressures and temperatures, the characteristics of the basic steam cycle are the same.

EFFECTS OF AGE

The effects of age on a ship should be considered with an appreciation for the fact that there is a vital difference between the routine demands of peacetime operations in the years preceding World War II and the stress of cold-war or combat operations.

Regardless of the degree of operational stress, it is the shipboard engineer's job to keep breakdown to a minimum by maintaining constant vigilance for the telltale signs of an imminent machinery casualty.

As the age of a power plant increases, the engineer must live with an air of expectancy and be ready to prevent, rather than repair, casualties. The tempo of fleet operations leaves little time for in-port maintenance or repair, lack of experienced engineering petty officers prolongs downtime, and lack of funds prevents the purchase of needed tools. In addition, repair parts for old machinery become increasingly difficult to obtain, as manufacturers cease automatically to produce replacement parts for their overage equipment. Thus, an engineer may have to wait from four weeks to six months for a replacement part to be manufactured. So, the watchword of the engineer must be *foresightedness.*

Engineering aboard ship is not a nine-to-five job. It is a way of life. Frequent inspection and constant vigilance, coupled with alert interpretation of properly maintained engineering records and logs are the keys to keeping a ship operating. The results of these practices might not be visible to the uninitiated, but a ship so served will possess a functioning firemain, reasonable watertight integrity, satisfactory damage-control posture, and all the other attributes necessary for meeting an enemy under combat conditions.

PERSONNEL PROBLEMS

For years, men who have steamed the Navy's conventionally powered ships have been affectionately referred to by their shipmates as the "black gang" or "snipes." But, there are times when these terms might not be used with affection or in jest, and to talk of a man who does not perform satisfactorily in a more prestigious billet "being banished" to the Engineering Department does not help to attract topnotch engineers. Some general line officers even fear that if they are designated engineering subspecialists, they will continue to be assigned to what they consider unglamorous billets and that their opportunities for command at sea and promotion will thereby be reduced. Fortunately, these philosophies are the exception and not the rule.

In actuality, the most devastating factor against attracting and retaining good enlisted personnel in the engineering ratings reduces simply to "poor working conditions." Characteristic of these conditions are long watches in hot and humid machinery spaces, often with reduced manpower allowance, followed, all too frequently, by periods of urgent maintenance and repairs which by their very nature can only be accomplished in port. So, while the rest of the crew goes ashore, the engineers may have to stay aboard to get the ship ready once again for sea, or merely to fulfill in-port watch requirements. These conditions, aggravated by edicts that the ship must be underway on, for example, two hours' notice, often throw a disproportionately heavy work load on the engineers. Yet, therein lies the challenge to all who traditionally work in the dirty shirts and dungarees. For, with this work load, grows a pride of accomplishment in having a knowledge that the ship is ready to meet all her commitments on schedule. And a justified pride this is, since almost every shipboard evolution, every piece of equipment involves, in one way or another, the Engineering Department: main propulsion, laundry equipment, fresh-water production, galley steam, sound-powered communications, refrigeration, air-conditioning, lighting, heating, ventilation, movies and entertainment system, liberty-boat engines, damage control, weapons-system power, and so forth and so on. The engineer provides the very source of life to a ship.

THE FUTURE

Each engineer aboard ship holds the key to the future. As he serves aboard various ships of the fleet and performs his duties diligently, he must also try to improve the fleet. He should be able, by utilizing procedures contained in Navy directives to submit his comments through proper channels, to assist the fleet by:

1. impressing upon competent authority the need for an accelerated shipbuilding program;
2. serving to bring about a re-evaluation of the Navy's admirable, but sometimes overworked, "can do" attitude;
3. stressing throughout the command structure the overwhelming advantage of long-range combat effectiveness over short-term expediency;
4. stressing the need for a realistic overhaul schedule;
5. stressing the need for thorough, comprehensive overhauls with proper funding by higher authority;
6. emphasizing the need for expanded training facilities for fleet engineering afloat and for making such schooling a requirement for junior officers before they are assigned to shipboard engineering duty.

The chapters that follow are at times presented with a touch of philosophy by those who have seen the effect of improper shipboard actions. For the most part, however, the *Naval Engineer's Guide* has been prepared in order to provide engineering personnel with a ready-reference outline for the accomplishment of numerous engineering evolutions, thereby aiding them in the performance of their work.

CHAPTER 2

Major Equipment in Marine Power Plants

The purpose of this chapter is to provide the reader with a rapid review of major steam power plant equipment functions, operating procedures, maintenance requirements, and safety procedures. It must be understood, however, that this book is not intended to serve as a guide for working applied thermodynamic problems or as a design text for future power plant innovations. Although it is assumed that the reader has been exposed to at least the general arrangements of a steam shipboard power plant, it would be unrealistic to prepare a guide of this nature without discussing the major equipment installed. System arrangements, however, vary from ship to ship. Further guidance in this area can be found in *Principles of Naval Engineering* (NavPers 10788B) and in various marine engineering textbooks. This is not to imply that book work can take the place of getting into a set of coveralls and personally tracing out a ship's system. Any young operating engineer or experienced hand who traces out the systems of his ship is certain to achieve a thorough knowledge of her plant arrangement. Ships, like people, are individuals and must be studied on that basis if their most intricate workings are to be understood.

It may be helpful at this point to quote from *Principles of Naval Engineering* on the subject of terminology:

> The boilers in a propulsion plant may be identified as *propulsion* boilers (or occasionally as *main* boilers) when it is necessary to distinguish between propulsion boilers and the auxiliary boilers that are installed on some ships. The turbines are identified as propulsion turbines when it is necessary to distinguish between them and the many auxiliary turbines that are used on all steam-driven ships to drive pumps, forced draft blowers, and other auxiliary units. The propulsion turbines are also sometimes referred to as the *main engines,* although this usage is not considered particularly desirable. The term *propulsion unit* is correctly used to identify the combination of propulsion turbines, main reduction gears, and main condenser in any one propulsion plant; however, the term *propulsion unit* may also be used in a more general sense to indicate any major unit in the propulsion plant.
>
> The propulsion machinery spaces may be physically arranged in several ways. Some ships have *firerooms,* containing boilers and the stations for operating them, and *enginerooms,* containing propulsion turbines and the stations for operating them. On some ships, one fireroom serves one engineroom; on others, two firerooms serve one engineroom. Instead of firerooms and enginerooms, many large ships of recent design have spaces which are called *machinery rooms.* Each machinery room contains both the boilers and the propulsion turbines that serve a particular shaft. On some recent ships that have certain automatic controls, the propulsion machinery is mainly operated from separate enclosed control stations located within the machinery room.

No matter what arrangement of machinery spaces is used, the propulsion machinery is usually on two levels. The condensers and the main reduction gears are on the lower level. The propulsion turbines and the high speed pinion gears to which they are connected are on the upper level, with the low pressure turbine exhaust directly over the condenser. The boilers occupy space in both the lower level and the upper level areas; the stations for firing the boilers (sometimes referred to as "the firing aisle") are on the lower level, while the stations for operating the valves that admit feedwater to the boilers are on the upper level. The upper and lower levels are normally connected through a series of catwalks, platforms, and ladders. The boilers are usually located on the centerline of the ship or else they are distributed symmetrically about the centerline. The long axis of the boiler steam drums runs fore and aft rather than athwartship. Other machinery, including the propulsion auxiliaries, is arranged in various ways depending on space and weight design considerations. [*NavPers 10788A.*]

ARRANGEMENT OF PROPULSION MACHINERY

To understand the propulsion of a conventional steam-driven ship, it is necessary to visualize the general configuration of the entire plant and to be conversant with the physical and energy relationships in the basic propulsion cycle. Figure 1 is a flow schematic that depicts the energy relationships in the basic cycle of a conventional steam-driven ship. Figure 2 is a flow schematic that depicts the basic major components of the propulsion plant of a typical U. S. Navy combatant ship. The thermodynamic design of the components, which are connected by appropriate piping systems, allows them to interact as a cohesive and integrated steam power plant.

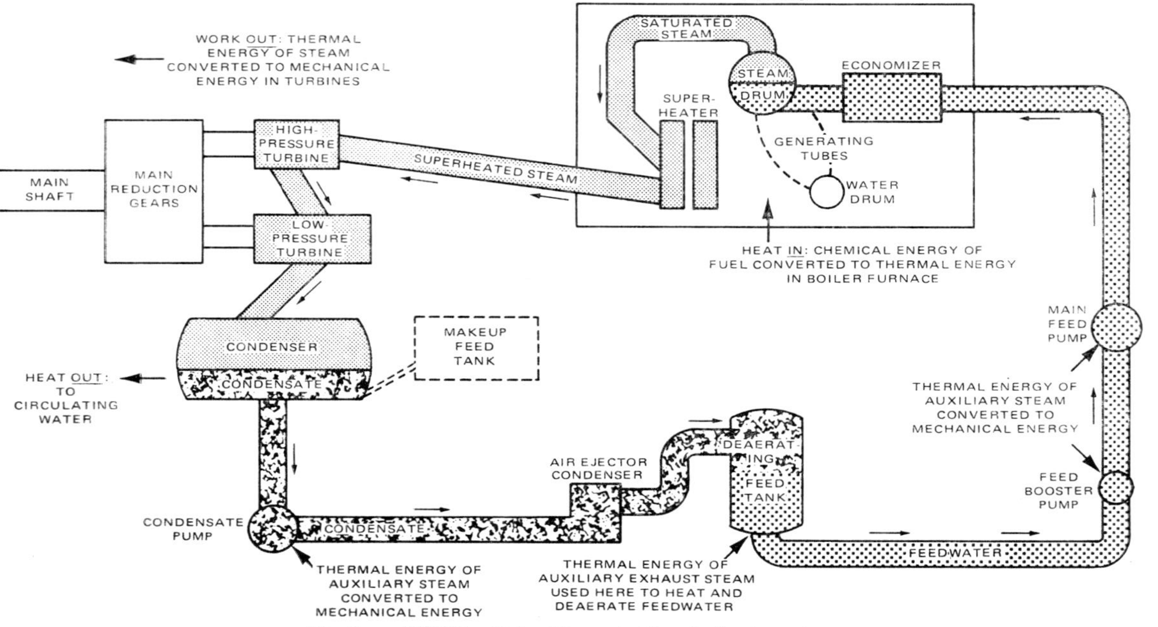

NavPers 10788B, *Principles of Naval Engineering*

Figure 1—Energy relationships in the basic propulsion cycle of a conventional steam-driven ship

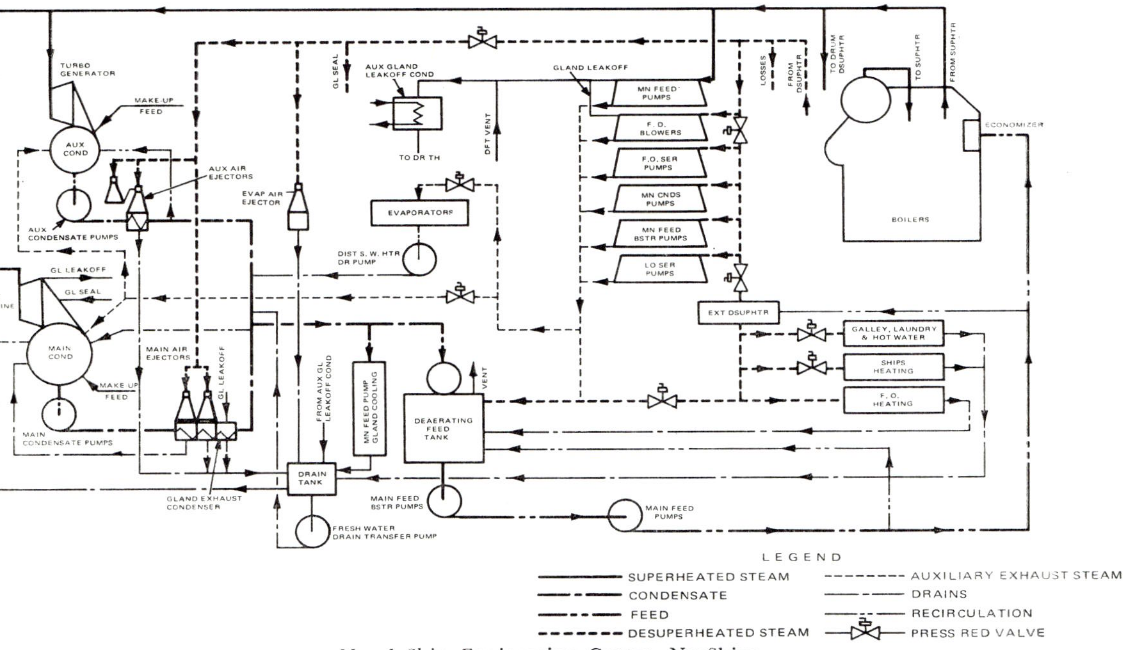

Naval Ship Engineering Center, NavShips

Figure 2—Flow schematic of major components of a typical U. S. Navy combatant ship propulsion plant

Propulsion machinery, which on most steam-driven surface ships is located amidships,* consists of:

Propulsion boilers
Propulsion turbines
Condensers
Reduction gears
Pumps
Forced-draft blowers
Deaerating feed tanks, and
Other auxiliary machinery units that directly serve the major propulsion units.

Turbo-generators and their auxiliary condensers are usually located in the propulsion machinery spaces; other engineering equipment that is not directly associated with the operation of the major propulsion units is located in or near the propulsion machinery spaces or in other parts of the ship, according to the space available.

BASIC STEAM-CYCLE COMPONENTS

As stated previously, the purpose of this chapter is to provide the reader with a rapid review of equipment functions and operation, maintenance, and safety procedures related to the basic shipboard engineering steam cycle. Such an arrangement as might be found on a steam-powered naval ship is shown in Figure 2. The major components of the basic steam-propulsion system will now be discussed in detail in terms of general requirements, inspections, safety precautions, and maintenance and operation.

---

* Certain amphibious ships, tankers, and auxiliary types may have their propulsion system located in the stern area.

BOILER

In a conventional steam-turbine propulsion plant, the boiler is the heat source, or high temperature region, of the thermodynamic cycle. Practically all boilers used in propulsion plants are designed to produce both saturated steam and superheated steam. Details for a specific ship's boiler can be found in the manufacturer's instruction booklet, and general reference material can be found in NavShips *Technical Manual,* Chapter 9510.

Some of the major boiler subsystems are the:

1. Forced-Draft Blower, which furnishes air under pressure to the boiler;
2. Main Steam Stop Valve, which prevents the leakage of steam from the boiler to the main turbine;
3. Blowdown Valves, which provide the means of freeing the boiler of dirt and grease, and of reducing the salinity of the boiler feedwater;
4. Soot Blowers, which provide the means of freeing the boiler fireside and superheater tubes of deposits of soot and film;
5. Air Registers, which provide the means of mixing air and fuel for proper combustion;
6. Safety Valves, which release excessive steam-drum pressure;
7. Fuel Atomizers, which provide proper fuel atomization;
8. Economizer, which is used to raise the temperature of the incoming boiler feedwater;
9. Guarding Valves, which, in new boilers, are usually installed in the blowdown system. These

valves are usually located in the common blowdown line from each boiler, between the boiler blowdown valves and the overboard discharge valves, to provide additional insurance against steam leakage from the boiler and to prevent salt water from entering a boiler that is secured or open for repairs.

General Requirements

It is the engineer officer's duty to be fully acquainted with the boiler system and, by periodic inspection, to satisfy himself that the

1. boiler is clean.
2. furnace lining protects casing, drums, and header.
3. pressure parts are being maintained.
4. sliding feet are operable.
5. operating condition of burners, safety valves, operating instruments and other hardware is satisfactory.
6. idle boilers are secured at all times.
7. fuel oil is free of saltwater.
8. feedwater is free of sea salts and entrained oxygen.
9. covers are on non-steaming stacks.

Inspections

Whenever parts are exposed for cleaning and overhaul, inspections should be conducted and results entered in the Engineering Log and the Boiler Record. Unusual circumstances, details of casualties sustained, remedies applied, and causes, if determinable, should be reported to the type commander (copy to Nav-Ships).

Safety Precautions

When draining boiler watersides and when boiler watersides and firesides are open, the following precautions should be taken:

1. *Emptying boiler*

   Never empty boiler to overboard through the bottom blow sea valve. Empty by

   a. pumping down with fire and bilge pump or portable pump connected by hose to the hose connection in bottom blow line. Open drains during pumping.

   b. draining to bilges and then operating bilge pump to discharge overboard. Feedwater really cleans bilges, so does old foam normally used in firefighting.

2. *Before opening boiler*

   a. make sure that drains, vents on drums, and header are open before loosening manhole plates and covers.

   b. stand clear when opening.

3. *When boiler is open*

   a. air out firesides and watersides before sending men into these work areas, because of the possible presence of inflammable or noxious vapors.

   b. do not use naked electric lights in an open boiler. This is prohibited. Portable sealed lights may be used, but battery-powered flashlights are preferred.

   c. close, log, and tag all valves that might permit

steam or water to flow into boiler. Lock remote valves topside.

d. have a man outside in communication with men in boilers.

e. do not allow any work in dead fireroom unless adequate ventilation is provided.

f. take special precautions to provide adequate ventilation if boilers in fireroom are connected to a stack that is connected to steaming boilers.

Maintenance and Operation

1. *Working on boilers*

   Methods for laying up boilers, taking care of firesides and watersides, setting up manhole and manhold plates and seats, and so forth, are provided in type commanders' instructions, the boiler instruction book, and NavShips *Technical Manual.*

2. *Frequency for cleaning boilers*

   Details are described in NavShips *Technical Manual,* Chapter 9510. However, these guidelines can be superseded by type commanders' instructions. In general,

   a. firesides should be cleaned every 600 hours.

   b. watersides should be cleaned every 1,800 to 2,000 hours, unless waived by type commander.

3. *Hydrostatic testing*

   Methods for testing are described in detail in NavShips *Technical Manual,* Chapter 9510.

4. *Fuel-oil burners*

   Oil burners provide effective fuel-oil burning

by mechanical pressure atomization. It is highly important that

a. fuel oil temperature and pressure be correct. The larger the sprayer plate size, the greater its importance. The higher the sprayer number, the smaller its opening.
b. proper sprayer plate sizes be used at all times.

5. *Blowing tubes*
   a. Soot is a very effective heat insulator. If a boiler is to generate steam efficiently, soot deposits must be thoroughly removed from its firesides at frequent intervals. This applies especially to high-rate express type boilers. Soot deposits that are permitted to remain on top of the lower drums or between the tubes will absorb moisture from the air. Moisture will produce sulphuric acid from elements contained in the soot and this acid is highly corrosive when it comes in contact with the boiler metal. Furthermore, undisturbed soot deposits can, in a comparatively short time, pack into a mass, too solid to be removed by soot blowers, and must then be removed by tedious hand cleaning. Where boilers are fitted with permanently installed soot blowers, the following minimum requirements as to their use should be met:

      (1) While underway, blow tubes in all steaming boilers at least twice each day at intervals of as nearly 12 hours as practical.

(2) While at anchor, blow tubes in all steaming boilers at least once each day.

b. When practical, blow tubes immediately after the making of heavy smoke from any cause (lighting off or smoke screens, for example).

(1) Minimum requirements for frequency of soot blowing are given in paragraph 5a (1) and (2). However, tests conducted under closely controlled conditions show that, for most economical operation, the soot blowers should be used once each 4-hour watch underway and twice each day in port.

(2) When blowing tubes, due consideration must be given to the effect on the upper decks. The officer of the deck must be consulted before the operation is undertaken. It should be noted that the more frequent the use of soot blowers, the less smoke and soot will be deposited topside.

(3) Steam soot blower elements should be rotated at a rate of 10 to 15 seconds per 90 degrees of rotation, making from 1 to 3 complete traverses with each element, depending upon the condition of boiler firesides.

(4) Whenever practical, shortly before burners are secured, the firesides (tubes, baffles) of the boiler should be cleaned of accumulated soot through use of steam blowers, if installed. A high air

pressure will materially assist in the cleaning. Steam soot blowers should be used only while fires are lighted in the boilers.

c. Sequence for blowing tubes—In order to ensure maximum efficiency of the soot blowers, the uppermost row of economizer tubes, or air-heater tubes, when installed, should be cleaned of all soot deposits. This should be done during the regularly scheduled boiler fireside cleaning periods, required every 600 hours. Once this row of tubes has been cleaned and the passage cleared, the sequence of soot-blower operation should be changed to blow the top soot blowers first. This will clear the gas passage of any soot that has collected since the last blowing of tubes or mechanical cleaning. After blowing the top soot blowers, the sequence outlined in the boiler technical manuals should be followed. The uppermost soot blowers, economizer or air heater, as applicable, are used at the beginning and at the end of the sequence.

6. *Boiler Blowdown Procedures*

Details for proper blowdown procedures are provided in NavShips *Technical Manual,* Chapter 9510. It is extremely important that these be faithfully followed.

7. *Boiler Light-off Procedures*

Details of the procedures for boiler light-off are given in NavShips *Technical Manual.*

a. Chapter 9510 of the *Manual* provides

(1) general steps in lighting-off both open

boilers and full boilers when steam is available and when it is not; and

(2) additional instructions for integral superheaters not separately fired and for divided furnace boilers.

b. Under ordinary circumstances, bring steam up in 2 hours. In emergency, bring it up in 30 minutes, if there is no new boiler brickwork.

c. Color of oil flame

(1) incandescent white flame with furnace walls clearly discernible indicates excess air.

(2) as air is reduced, flame color becomes pale yellow, then yellow-orange and then orange-red.

(3) on a well-designed installation operating on minimum excess air, the end of the flame farthest from the atomizer should be yellowish-orange or golden.

8. *Casualties*

The following brief summary should help in identifying or avoiding casualties:

a. Carbon on atomizer tips—improper location of tips.

b. Carbon on burner openings—atomizers drawn back too far.

c. Carbon on deck—improper atomization, usually due to improper tip location but sometimes due to excess air entering around cone of oil.

d. Flarebacks—pressure in boiler greater than in fireroom due to explosion of oil vapor mixture or to a drop in fireroom air pressure. This normally occurs in light-off or in attempting to relight a burner off a hot wall. Never do the latter—use a torch.

e. Oil fires—oil can catch fire in accumulations throughout space. Don't let it accumulate.
f. Oil in boilerwater and feedwater—check drains that can be contaminated, such as those located on fuel oil heat exchangers, and lub oil heat exchangers.
g. Choked atomizer—plugged atomizer or water in fuel oil puts out fire.
h. Atomizers sputter—loss of fuel oil suction (bottom of suction tank).
i. High water—water too high in the steam drum —priming occurs.
j. Low water—water too low in the steam drum—very serious—usually results from a failure of the feedwater regulator, a main feed pump casualty, or a major leak in the boiler or feed piping systems.
k. Serious steam leak or ruptured tube—details in NavShips *Technical Manual,* Chapter 9510.

Note: A detailed summary of Inspections, Tests, Boiler-cleaning Requirements, Records and Reports, and Safety Precautions can be found in Appendix A.

MAIN PROPULSION TURBINE

In general, the main propulsion turbine is used to convert the thermal energy of the steam leaving the boiler into mechanical energy. In effect, the turbine uses the high velocity of the steam to impart rotation to the main propulsion turbine rotor which, in turn, rotates reduction gear shafting and finally imparts low-speed propeller shaft rotation to the ship's propeller.

Turbine installations vary to such an extent that detailed instructions cannot be given in any one text or manual. NavShips *Technical Manual* gives sufficient information relating to general principles. The instruction book that goes with each installation gives specific details.

General Requirements

It is the engineer officer's duty to be fully acquainted with the turbine system and, by periodic inspection, to satisfy himself that

1. turbine bearing temperatures are correct and other turbine hardware is functioning properly.
2. warm-up, lighting-off, and securing procedures are properly followed.
3. he is familiar with the taking of leads on bearings, the taking of gage readings, and the positioning and axial adjustment of thrust bearings.
4. he is familiar with required clearance limits for each bearing.

Inspections

A record of all inspections should be maintained by appropriate entries in the Engineering Log.

1. *While underway:*

WHEN MADE	ACTION	PURPOSE
a. at very short intervals during each watch	inspect bearings	to detect signs of overheating

WHEN MADE	ACTION	PURPOSE
b. at very short intervals during each watch	check for flow of oil through sight glasses	to check oil flow
c. at very short intervals during each watch	operate finger piece and telltale clearance	to check axial clearance
d. at very short intervals during each watch	inspect pressure gages	to check pressures
e. once a watch	take dummy clearance	to check axial clearance
f. once a watch	check ball thrust bearings and radial bearings	listen to detect defects in or damage to bearings
g. daily	check parts under vacuum	for air leaks

2. *While at anchor or in port*:*

WHEN MADE	ACTION	PURPOSE
a. daily	jack turbines 1¼ turns	to prevent freezing
b. daily	circulate lubricating oil in system	to prevent accumulation of water and sediment in pockets and to form oil film while jacking

* These requirements sometimes vary with the PMS schedule for the ship, if installed.

WHEN MADE	ACTION	PURPOSE
c. daily	run air ejector one-half hour, or longer if unusual conditions, such as leaky stop valves, require	to dry out turbines
d. daily	operate lubricating oil separator	to prevent water getting mixed with oil when sent to the bearings
e. weekly	check valves, cocks, joints of steam, exhaust, and drain lines	for tightness
f. weekly	operate and oil throttle valves	to prevent sticking
g. weekly	operate and oil, if possible, all valves not used	to prevent freezing of parts
h. quarterly	inspect exhaust trunk and last few stages of low-pressure turbine through manhole plate	to detect corrosion or other defects
i. quarterly	sound with hammer, holding down bolts, ties, and chocks of turbines	to detect signs of loosening turbine fastenings
j. quarterly	clean the steam strainers (clean oftener if extensive cruising has been done)	to prevent foreign matter entering turbine, and to insure integrity of the strainers

WHEN MADE	ACTION	PURPOSE
k. quarterly	remove turbine inspection plates	to determine existence of loose blades or shrouding, and to look for corrosion
l. quarterly	check thrust shoes for clearance and conditions of bearing surface	to ensure proper position of rotor
m. quarterly	blow out thrust bearing with air after examination	to prevent foreign matter remaining
n. quarterly	check main bearings for clearance condition of journal, and bearing surface	to ensure radial clearance is correct
o. quarterly	check exhaust trunk	for tightness, evidence of corrosion, and loose bolts
p. quarterly	test and set steam relief and sentinel valves	to ensure proper protection against excessive pressures
q. quarterly	check location, tightness, and condition of rotor balance weights through inspection openings	to detect corrosion, erosion, or other defects
r. annually	calibrate gages	to ensure accurate readings

WHEN MADE	ACTION	PURPOSE
s. before getting underway	inspect sliding feet of turbine, if fitted	to see that they are not stuck
t. before getting underway (cold)	check dummy clearance	to check axial clearance
u. before getting underway (hot)	check dummy clearance	to check axial clearance

3. *Shipyard overhaul:*

Open up and examine all thrust bearings and radial bearings.

Safety Precautions

1. Do not use auxiliary exhaust steam for warming up the main turbines (Use of Exhaust Steam).
2. Be sure that the lubrication system is in operation before turning over the main engines (Precautions Before Turning Turbine Over).
3. Keep the main turbine rotors turning over continuously while warming up the main engines (Method of Warm-Up).
4. Do not put way on the ship when turning over the main engines (Warming up by Spinning).
5. Never fail to investigate any noise emanating from a turbine (Rubbing and Other Unusual Noise in a Turbine).
6. If turbine vibrates, slow down, investigate, and endeavor to locate the cause (Turbine Vibration).
7. Except in an emergency, do not admit steam

into an astern turbine until steam has been shut off from the ahead turbine, and vice versa (Additional Precautions with Throttle Valve).

8. In getting underway, be sure that all lines are properly drained, in order to prevent danger from "water hammer" (Warming Up Main Steam Line).
9. When steam pressure drops, do not open the main steam throttle valve to such an extent that the working pressure of the steam is brought to a dangerously low point (Handling the Steam).
10. Stop engines if oil supply fails (Failure of Oil Supply).
11. If throttle valve sticks open, close bulkhead stop as soon as possible (Throttle Sticks).
12. Extreme care must be taken to prevent the entry of foreign matter into the turbine when it is opened for inspection (Precautions to Prevent Entry of Foreign Matter).
13. About 24 hours after securing and when turbine is thoroughly cooled, close turbine drains (Securing Main Engines). [Excerpted from Nav Ships *Technical Manual,* Chapter 9410.]

Maintenance and Operation

1. *Warm-up*

Turbines must be warmed up during the lighting-off procedure, because all parts of a turbine must expand at the same rate as the surrounding installation. If this is not done, damage due to distortion can occur. If the rotor and casing are not evenly heated, their unequal expansion will result in distortion of the rotor or casing, or both.

2. *Lighting-off and securing a turbine installation*

   This is covered in NavShips *Technical Manual,* Chapter 9410. Similar instructions are also contained in type commanders' Lighting-off and Securing sheets. These sheets should be a part of the Engineering Department Standing Orders, and engineering personnel should be required to use them (*See* Appendix B).

3. *General instructions*

   General instructions for other features of operation are contained in NavShips *Technical Manual,* Chapter 9410, Part 3, which covers such features as:

   a. Standing by indefinitely
   b. Standing by on "½-hour notice"
   c. Getting ready to get underway
   d. Setting standard speed.

4. *Importance of high vacuum*

   A low exhaust pressure (high vacuum) is important because it permits a greater range of expansion for the steam, and thus makes a greater amount of Btu/lb steam available for useful work. In order to obtain the greatest operating economy, the optimum vacuum for which the turbine was designed, must be maintained by:

   a. ensuring that astern turbines do not leak;
   b. keeping gland packing in good condition;
   c. maintaining a steam pressure of between ½ psi and 2 psi gage (psig) on the gland seal system when turbines are in operation;
   d. ensuring that there are no air leaks in

condenser, exhaust trunks, throttles, lines to air ejectors, gage lines, idle condensate pump-packing valves, make-up feed lines, etc.;

e. maintaining adequate water in the feed tank on which vacuum drag is being taken.

5. *Purpose of cruising combinations*

A combatant ship operates most of the time at speeds far below maximum, and when doing so requires only a fraction of the power for which her main turbines were designed. Operating economy is attained by the use of cruising turbines specially designed for operating at reduced speeds. The cruising combination should be used when not maneuvering and when standard speed is such that the cruising combination can furnish the required revolutions per minute (rpm).

6. *Casualties to turbines*

Casualties to turbine installations are covered in NavShips *Technical Manual.* Chapter 9410 provides details for handling the following casualties:

a. Vibration of and unusual noise in turbines
b. Difficulty in turning
c. Failure of oil supply
d. Large loss of oil
e. Loss of vacuum
f. Stuck throttle
g. Failure of circulating pump
h. Locked shaft

7. *Class details*

Details relating to ship class in regard to turbine rpm, astern power conditions, major turbine adjustments, and repairs and principles of Kingsbury thrust bearings can be found in NavShips *Technical Manual,* Chapter 9410, Section I.

REDUCTION GEARS

The reduction gears convert the high-shaft rpm produced by the turbine to a low propeller shaft rpm for developing a proper propeller rpm for desired thrust. This permits the turbine prime mover and the propeller to operate at their most efficient speeds. Since reduction-gear installations vary with each class ship, instruction books should be consulted for details of installation.

General Requirements

It is the engineer officer's duty to be fully acquainted with the reduction-gear system and to ensure that

1. no inspection plate, connection, fitting, or cover which permits access to the gear-casing is ever removed without his specific authority.
2. when gear cases are open, precautions are taken to prevent the entry of foreign matter. Openings should never be left unattended unless satisfactory temporary closures have been installed. Before closing gear case, the engineer officer must make inspections as required. (NavShips *Technical Manual,* Chapter 9420.)

Inspections

The following tests and inspections of reduction gears should be made:

1. *Daily,* when at anchor

   Jack gears so that main gear shaft is moved about $1\frac{1}{4}$ turns. This jacking should be done with lubricating oil circulating in the system.

2. *Weekly*

   When oil sump is located at the skin of the ship, sight gears for signs of pitting due to salt-water leakage into lubricating system.

3. *Quarterly*

   a. Sound with hammer, all hold-down bolts, ties, and chocks to detect signs of loosening of casing fastenings.

   b. Remove inspection plates and inspect gears. Wipe off oil at different points and note whether the surface is bright, or, if already corroded, whether new areas have been affected.

   c. Jack main gear fore and aft to measure propeller-thrust clearance. Record the clearance.

   d. Inspect oil-spray nozzles. If the observed spray pattern indicates that they are not functioning properly, clean spray fittings.

4. *Semiannually*

   Inspect couplings with self-contained lubricant; check quantity of lubricant, and replenish as required.

5. *Annually*

   Inspect all turbine couplings; remove sludge, check backlash and stone the tooth surfaces, if necessary.

6. *During shipyard overhaul*

   a. Inspect condition and clearances of thrust shoes to ensure proper position of gears. After the inspection, blow out thrust shoes with dry air. Record the readings. Inspect thrust collar nut and locking device.

b. If turbine coupling inspection has indicated undue wear, check alinement between pinions and turbines.

c. Clean oil sump.

7. *Every seven years*

a. Inspect clearance of bearings and journals and the condition of the rubbing surfaces. Record the bearing-crown thickness or lead readings of all gear-wheel and pinion bearings.

b. Take and record alinement readings.

c. NavShip authority for lifting gear-case covers is not required. Covers should be lifted when trouble is suspected. An open gear case is, however, a serious hazard to the main plant. Rags in oil lines and tools in gear teeth have caused casualties attributable directly to gear cases having been lifted. The dangers of uncovering the gears must be carefully weighed against the seriousness of the suspected internal trouble before deciding to lift a gear-case cover. The seven-year interval may be extended by the type commander if operating conditions indicate that a longer interval between inspections is desirable.

8. *Scheduled trials*

a. Before Trials

Conscientious observance of the instructions in NavShips *Technical Manual,* Chapter 9420, and the correction of defects disclosed by regular tests and inspections listed should assure that the reduction gears are ready for full power at all times. The general opening-up of gear cases, bearings, thrusts, etc., immediately before trials is to be discouraged, because results could be harmful rather than beneficial. In addition to inspections which may be directed by proper authority, the inspection plates should be removed, and the tooth contact and the condition of the teeth examined. Bluing an area of the gear will help in determining tooth wear after trials.

b. After Trials

In addition to the inspection which may be directed by proper authority, remove the inspection plates and examine the tooth contact and the condition of the teeth to note changes that occurred during the trial. Check the clearance of the Kingsbury thrust bearing. Running at high power will, in a few hours, show improper contact or abnormal wear that would not show up in months of running at lower powers.

Safety Precautions

1. In case of churning or emulsification of the oil in the gear case, the gear must be slowed or stopped until the defect is remedied.
2. If for any reason, the supply of lubricating oil to the gears fails, the gears should be stopped until the cause can be located and remedied. The gears should not be operated until the normal supply of oil has been restored.
3. When bearings are known to have been over-

heated, gears should not be operated (except in cases of extreme emergency) until bearings have been examined and defects remedied.

4. Temperature of oil from a bearing, as measured by the thermometer, should not exceed 180° F. maximum or a 50° F. rise above cooler-outlet temperature, whichever is the lesser (unless NavShips has specifically approved an exception).
5. When salt water is found in a reduction gear system, immediate corrective action is required to locate and seal off the source of the water. Remove contaminated oil and flush the system.
6. Naked lights should be kept away from vents while gears are in use, because the oil vapor may be explosive.

Maintenance and Operation

1. *General*
   a. Lubricating oil must be furnished to the propulsion machinery of a locked shaft or a trailed shaft when a ship is underway.
   b. When the turning gear is engaged to lock a shaft, the throttles should be secured by a chain and lock. The custody of this key is the responsibility of the engineer officer.
   c. The use of pinch bars near the ends of teeth for moving the pinions or gears to determine longitudinal movement is prohibited.
   d. If excessive flaking of metal from the gear teeth occurs, the gears should not be adjusted (except in case of emergency) until the cause has been determined. Care should be taken to

prevent the entry of the flakes into the general lubricating system.

e. Unusual noises should be investigated at once, and the gears should be operated with caution until the cause is discovered and remedied.

f. Inspection-plate joints should be kept free from paint.

g. Lifting devices should be inspected carefully before being used.

2. *Casualties*

The causes of casualties to the reduction gear and their remedies may be summarized as follows:

SYMPTOM	CAUSE	REMEDY
a. Wiping	Lack of oil	Depending on extent of wipe, replace bearing or clean by scraping. Restore oil supply.
	Poor bearing contact	Poor bearing contact usually results in localized wipes. Scraping to a mandrel to obtain contact across the entire length of the bearing is normally required. Severe overheating may damage a bearing sufficiently to require replacement.
	Pinched bearing	Pinched bearings require correct installation or, in the case of warped bearings, replacement.
	Misalinement of gearing	The wipe is due to the fact that the journal is not on

SYMPTOM	CAUSE	REMEDY
		the entire bearing length. Realinement of gearing will rectify.
	Insufficient clearance	Bore or scrape to correct clearance, or replace with bearing having correct clearance.
b. Wear	Normal	A small amount of wear is normal and no repair is required.
	Low oil viscosity	For high-wear rates, use of cooler oil or higher viscosity oil is required.
	Starts and stops	This wear is normal since the oil film is not present when the unit is at rest. When jacking, always supply oil to bearings, in order to reduce such wear.
c. Abrasion (circumferential scratches and embedded particles)	Vibration, dirt, or grit in oil	Locate source and correct. Clean lube oil system.
d. Corrosion	Acid or salt water in the oil, wrong type oil, or extreme pressure in the system	Replace oil.
e. Erosion	Oil stream	Very rare. Eliminate air in

SYMPTOM	CAUSE	REMEDY
		oil. Design changes may be required.
f. Cracking	Fatigue	Very rare in gear bearings. May be due to severe misalinement or heavy vibration. Locate source and correct.
	Poor bonding	Separation of babbitt from back caused by poor bonding. Bond failure can be detected by squeezing babbitt to see if oil squeezes out between babbitt and back, by dye penetrant method, or by ringing with a hammer. Cracks may be apparent on surface of the babbitt. Replace bearing.

MAIN CONDENSER

The thermodynamic function of the main condenser is to take the exhaust steam from the main propulsion turbine and change it into water by a condensing process. The condenser's function is to maintain a vacuum in the condensate system and to permit recovery of feedwater. Maintenance of this vacuum is essential for the efficient and safe operation of the main propulsion turbines. The type of condenser used in modern naval practice is a horizontal, single pass, surface-condensing type. Thus, seawater flows through all the condenser tubes in the same direction. The steam passing over these tubes is then condensed. A schematic of the main condenser system is presented in Figure 3.

General Requirements

It is the engineer officer's duty to ensure that

1. condensers are kept clean and tight.
2. idle condensers are kept either filled or completely dry.

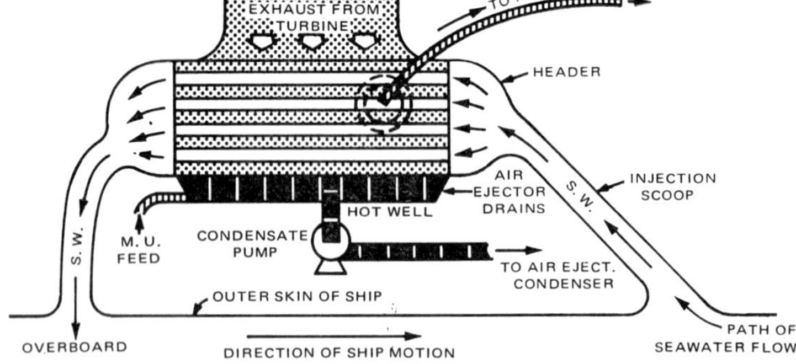

Division of Engineering and Weapons, U. S. Naval Academy

Figure 3—The Main Condenser

Inspections

The circulating waterside should be inspected and, if necessary, cleaned as follows:

1. Once each month, as permitted by ship's operations, and when zinc anodes are inspected or replaced.
2. As soon as possible after operation in polluted or shallow water.
3. During dry-docking or whenever the circulating water system is drained for other reasons.
4. Prior to initial sea trials and, if possible, prior to full-power trials.

Safety Precautions

NavShips *Technical Manual,* Chapter 9460, provides general safety precautions. When testing or cleaning condensers, the following precautions should be observed:

1. Close, tag, and wire shut all sea valves in the system.
2. Drain condenser before removing manhole plates.
3. Always replace manhole plates upon securing from work.
4. Pressure in condenser shells should not exceed 15 psig.
5. Do not be satisfied if one leak is found. Look for additional leaks.
6. Do not permit open flames in the vicinity of the open water chest—there may be pockets of hydrogen gas. Ventilate the chest thoroughly before allowing personnel to enter.
7. The seawater side of a condenser should be drained before the steam side is flooded and it must be kept free of water until the steam side is emptied.

Maintenance and Operation

1. *Circulating-water regulation*
   a. No throttling by any valve which is located less than 7 pipe diameters from the entrance to the water chest.
   b. Gate valve shall be kept at least 1/4 open to avoid flow erosion.
   c. Condenser water chest pressure shall not exceed 15 psig.

d. Relief valve provided on inlet water chest can serve as a sentinel valve. Its real purpose is to protect the condenser from excessive pressure caused by expansion of the water as it gradually warms to engineroom temperature.

e. Valves connected to *positive-displacement* circulating pumps shall not be throttled.

f. Sufficient water must be provided to the condenser to ensure that all tubes are completely filled with water.

2. *Procedures for Scoop Injection Operation*

   a. Under normal operating conditions: operate to maintain the maximum vacuum obtainable.

   b. For high-power operation/cold injection: restrict the circulating water flow at its discharge to allow only one inch of mercury vacuum higher than the designed vacuum.

   c. For cruising speeds

      (1) Normal injection temperature presents no problem. With cold injection temperature, extra high vacuum exists, and extra air leakage into the system results in a loss of economy.

      (2) It is good practice to throttle the flow of circulating water as necessary to limit the vacuum to about 0.2 inch mercury less than the maximum obtainable with full flow of circulating water, or as necessary to reduce the condensate depression to normal. If the condensate depression re-

mains between zero and about two degrees under low-power operation at medium injection temperature, (75°F.), it may be assumed that the air-removal equipment is removing air from the condenser, and it is not necessary to throttle circulating water.

3. *Repairs*

Detailed testing, cleaning, and repair procedures can be found in NavShips *Technical Manual,* Chapter 9460.

4. *Casualties*

The causes of casualties to condensers may be summarized as follows:

SYMPTOM	CAUSE
a. Inadequate vacuum	(1) Excessive air leakage into the vacuum system
	(2) Improper functioning of air-removal equipment
	(3) Improper drainage of condensate from condenser
	(4) Insufficient flow of circulating water
	(5) High injection temperature
b. High condensate depression	(1) Excessive air leakage
	(2) Improper condensate removal

SYMPTOM	CAUSE
	(3) Excessive flow of circulating water
	(4) Improper functioning of air-removal equipment
c. Improper condensate drainage	(1) Malfunctioning of condensate pump
d. Insufficient flow of circulating water	(1) Foreign matter clogging condenser tubes
	(2) Improper operation of circulating pumps
	(3) Injection and/or overboard discharge sea chests, strainers, piping, or valves may be obstructed
	(4) Air not venting from condenser properly
e. Condensate contamination	(1) Deterioration of the tube wall
	(2) Leaks between the tube sheet and the tube wall
	(3) Cracking of the tube wall
f. Large air leaks	(1) Loss of gland sealing steam pressure
	(2) Malfunction of the float control dumping valve for the fresh water drain collecting tank
	(3) Attempts to take on make-up feed from an empty reserve feed tank
	(4) Failure of steam supply to the air injectors

AIR EJECTORS

It has been noted that air entering the steam expansion and condenser system will collect in the condenser because that is where the lowest system pressure exists. This air contributes a partial pressure to the condenser. Therefore, air must be removed if the vacuum is to be maintained.

A steam air ejector is used to remove most of the air and uncondensed vapors from the condenser. Since a steam jet is used, provision must be made for condensing the expanding steam in order to conserve water. Consequently, the air ejector is designed to discharge into the shell of a small condenser (heat exchanger). Condensate formed therein is drained to the ship condensate system. A simplified system is provided in Figures 4a, 4b, and 4c.

General Requirements

It is the engineer officer's duty to ensure that he is fully acquainted with the operation of the air-ejector system.

Safety Precautions

1. Test sentinel valve at least quarterly.
2. Inspect steam strainers on a semiannual basis.
3. Always open the interstage valves before admitting steam to nozzles, and when lighting-off always close the steam supply valves tightly before closing interstage valves when securing.
4. Drain steam lines, open drain valves, and clear atmospheric vent line before lighting-off an air ejector.

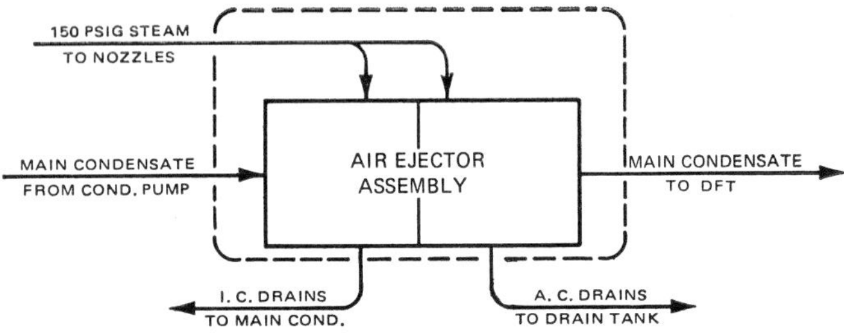

Figure 4a—Simplified System Schematic

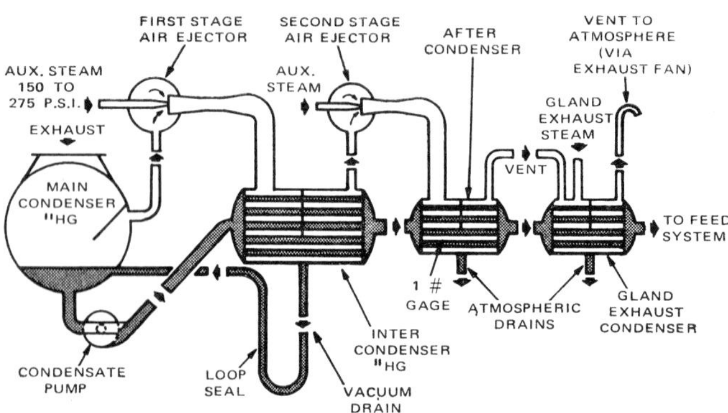

Figure 4b—Typical Two-Stage Air Ejector Assembly

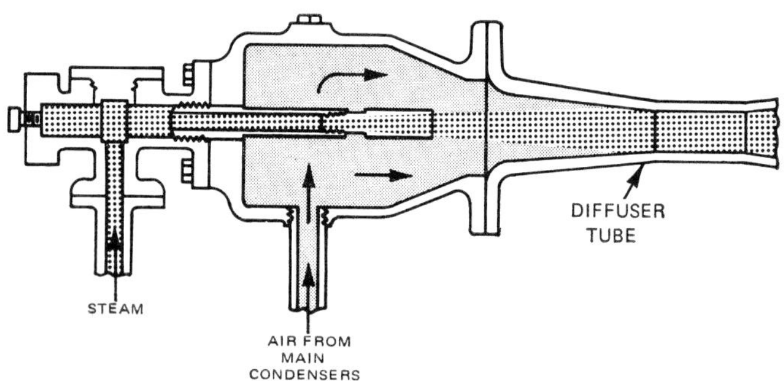

Figure 4c—Typical First-Stage Air Ejector

5. Always hydrostatically test the system after tube replacement or other major repairs.
6. Secure all isolating valves to avoid personnel burns when overhauling an operating air-ejector assembly.

Maintenance and Operation

1. *Light-off procedure*
   a. Drain steam lines.
   b. Start recirculating condensate (use main circulating pump).
   c. Open inter- and after-condenser drain valves.
   d. Check to ensure gage-line valve is open and that the atmospheric vent line is clear of condensate.
   e. Open first- and second-stage suction and discharge valves.
   f. Fully open the steam inlet valve to second stage. (Ensure design steam pressure is available. Value can be found on the nameplate data.)
   g. Fully open the first-stage steam inlet valve when condenser vacuum reaches 20″ Hg.
   h. When condenser vacuum reaches required vacuum, the air ejector is considered to be in full operation.
2. *Casualties*

   The causes of casualties to air ejectors can be summarized as follows:

SYMPTOM	CAUSE
a. Failure to maintain condenser vacuum	(1) Improper functioning of steam-reducing valve, which will occur when steam pressure is too low

SYMPTOM	CAUSE
	(2) Clogged steam strainers
	(3) Insufficient cooling water circulation in condenser units
	(4) Insufficient flow of main condensate due to faulty spring-loaded unloading valve
	(5) Leaking sentinel valve
b. Erratic operation and loss of vacuum	(1) Tube leaks
	(2) Dirty condensers
	(3) Eroded nozzles
	(4) Improper repair with poor-fitting replacement parts

DEAERATING FEED TANK (DFT)

Although undissolved air is removed from the condensate system by the air ejectors, a considerable amount of dissolved air remains in the condensate. Prior to using this water as boiler feed, the oxygen must be removed in order to prevent serious corrosion of the boiler tubes. The DFT removes the dissolved oxygen, heats the feedwater, and acts as a surge tank. Modern plant high temperature and pressure cause soluble or entrained oxygen to produce considerable corrosion. Carbon dioxide in air can combine with water to form carbonic acid which, in the presence of dissolved oxygen, makes the feedwater corrosive. This corrosive effect is accelerated as temperature and pressure increase. This problem is of prime importance today and is discussed in Chapter

9. Data regarding feedwater limits and definitions of terms relating to the subject are provided in Appendix C and the Glossary, respectively.

General Requirements

It is the engineer officer's duty to make himself thoroughly familiar with the operation of the DFT system. A schematic of a DFT is provided in Figure 5.

Maintenance and Operation

1. *Operating sequence*
   a. Water enters tank through condensate inlet header.
   b. It then passes through the preheater and vent condenser. Vapor and air pass around the preheater tubes on their way to be exhausted from the DFT, thereby heating the water in the tubes as well as accomplishing condensation of the vapor entrained in the venting air.
   c. From the preheater, the condensate passes into the water box.
   d. From the water box, fluid passes through a series of spring-loaded nozzles, so designed that when water is discharged under pressure, it is broken up into the dome of the tank where it is directed into the center of the tank by a circular baffle attached to the tank top. Thus, the partially heated water is broken up into a fine spray which causes a great amount of entrained oxygen and air to be released. After striking the baffle and the top of the tank, the

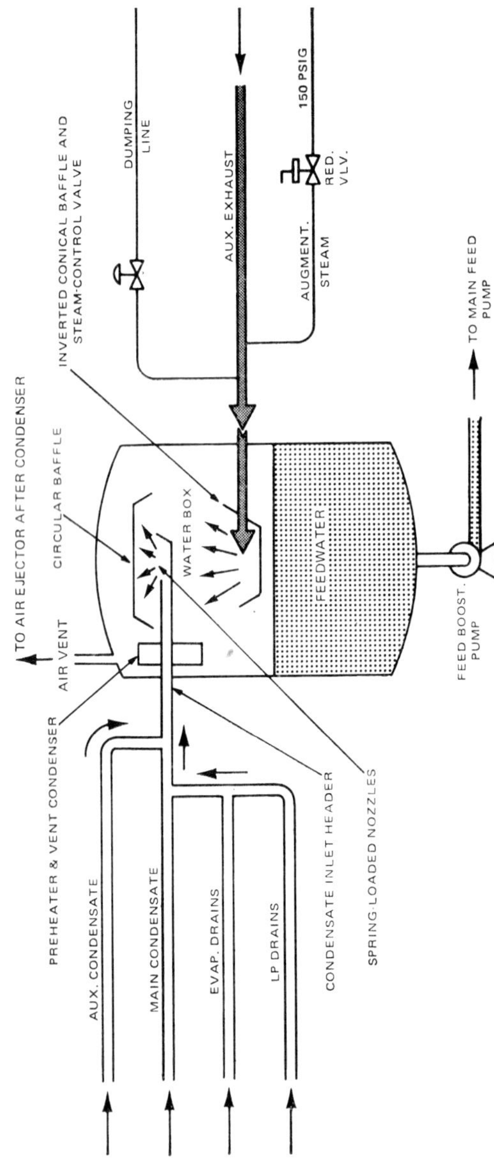

Figure 5—Deaerating Feed Tank

water drops downward and is caught by the inverted conical baffle (*see* Figure 5). It then flows down around the steam-control valve.

e. Steam enters the tank from the auxiliary exhaust line, passes through the ports of the steam-control valve, is sprayed outward, and deflected upward by the flanged disk of the valve.

f. The water dropping down into the conical baffle is picked up by this spray of steam and driven along with it. This causes the water to be heated and scrubbed by the steam, thereby releasing all the oxygen and air which has been left in solution with the water.

   This mixing of water and steam causes condensation of part of the steam which, with the water, falls to the bottom of the tank. The uncondensed steam, carrying the liberated air with it, rises in the tank outside the conical baffle and passes around the baffle and mixes with the water being discharged by the spray nozzles. Here, steam is mixed with the water spray and condensed.

g. The released air, plus a small amount of vapor, travels from the dome into the baffles in the preheater and vent condenser, where vapor is condensed and the air is exhausted from the tank through the vent box air vent.

h. The liquid on the bottom of the tank goes directly to the feed-booster pump suction.

2. *Casualties*

   Likely casualties to DFTs and their causes are:

SYMPTOM	CAUSE
a. Tank too full	(1) Malfunctioning feed-booster pump
	(2) Make-up feed valve open
b. Oxygen not being removed	(1) Faulty vent condenser
	(2) Defective spray valves

MAIN FEED PUMP

The requirement for the feed pump system is rather obvious. Pressure in the Deaerating Feed Tank is normally about 15 psig and boiler pressure is usually in excess of 600 psig. Therefore, there has to be some system that will raise the feedwater to a pressure higher than boiler pressure including overcoming the various piping and fitting pressure losses en route.

The method of developing the pressure is to install a small, low-speed main feed-booster pump downstream of the main feed pump. This booster pump maintains a positive suction head to the large, high-speed main feed pump in order to prevent flashing of the feedwater into steam. The main feed pump is usually a centrifugal pump which converts a high-velocity head to a high-pressure output.

General Requirements

1. Pump details vary and individual instruction books should be consulted for operating procedures.
2. General information on pump operation and maintenance can be found in NavShips *Technical Manual,* Chapter 9470.

Inspections

1. *Weekly*
   a. Operate all pumps by steam or power. If power is not available, turn over by hand.
   b. Lubricate the pressure-regulating governor, if required.
   c. Check condition of lubricating oil.
2. *Quarterly*
   a. Check thrust position of pump rotor.
   b. Sound and set up as necessary all foundation bolts. Secure all foundation dowel pins.
3. *Annually*
   Check bearing clearances by leads or crown thickness measurements.
4. *During Shipyard Overhaul*
   a. Check internal water-lubricated bearings, and shafts in way of same, for wear and scoring.
   b. Open pump and reduction gear casings; inspect and clean.
   c. Check clearances of all diaphragms and casing-throat bushings, impeller and casing wearing rings, pressure break-down drums and bushings, and renew as necessary.

Safety Precautions

1. Test relief valves, where fitted, to see that they function at designated pressure.
2. Never attempt to jack a pump by hand while the steam valve to the pump is open.
3. Do not use boiler feed pumps and other pumps in the feed system for purposes other than those

connected with boiler or feedwater service, except in an extreme emergency.

4. Do not tie down or otherwise render inoperative the overspeed trip, speed-limiting or speed-regulating governor.
5. Ensure that overspeed trips, where fitted, are set to shut off steam to the unit when rated speed is exceeded by 10 percent.
6. Ensure that speed-limiting and speed-regulating governors are set to limit the speed of the unit to the manufacturer's rated speed under rated conditions, and that rated speed is not exceeded by more than 5 percent for any condition of loading.
7. Check the setting of the overspeed trip and speed-limiting governors at least quarterly.
8. When inspecting or overhauling vertical rotary pumps *never* rely on a flexible coupling to support the total weight of the pump rotors. *Always* support the pump rotors by means of wire rope slings, block and tackle, or chocks before working on the pump.
9. Install recirculating lines and test valves to ensure the system is operable.

Maintenance and Operation

1. *Casualties*

The causes of casualties to the main feed pump can be summarized as follows:

SYMPTOM	CAUSE
a. Failure to deliver water	(1) Pump not primed

SYMPTOM	CAUSE
	(2) Pump not up to speed
	(3) Impeller plugged
	(4) Wrong rotation
b. Short in capacity	(1) Air leaks in suction or stuffing boxes
	(2) Pump not up to speed
	(3) Impeller partially plugged
	(4) Suction strainers fouled
	(5) Mechanical defects to wearing rings, impeller, or casing gaskets
c. Low pressure	(1) Pump not up to speed
	(2) Air leaks
	(3) Incorrect manifold discharge valves open
d. Pump loses capacity after starting	(1) Leaky suction line
	(2) Water seal plugged
	(3) Suction lift too high
	(4) Air or gases in water
	(5) Fouled strainer
e. Pump vibrates	(1) Misalinement
	(2) Warped foundation
	(3) Impeller partially clogged
	(4) Mechanical defects to rotating element: shaft binding, shaft bent, or bearings worn.

2. *Repairs*

   Repairs vary depending on pump type and manufacturer. Consult manufacturer's instruction booklet for specific pump installed.

TURBO-GENERATORS AND SWITCHBOARDS

The normal ship service generator is driven by a steam turbine. The machinery plant, in addition to producing shaft horsepower (shp) to drive the ship, must produce electrical power. Electricity provides the motivating energy for the ship's weapon and control systems, for the electrically driven auxiliary equipment, and for ensuring normal personnel comforts. Since the ship exists for the purpose of using her weapons, the electric plant's function is of primary importance. In most warships, the motive power for the generator is developed by a steam turbine, but gas turbines or diesel engines have also been used as prime movers.

General Requirements

1. Operating personnel should possess thorough knowledge of the electrical system aboard ship, including basic operation and maintenance procedures.
2. The general requirements for the turbo-generator prime mover are the same as those set forth for the main propulsion turbine on page 28.
3. Operating procedures should be obtained from manufacturer's instruction books, NavShips *Technical Manual,* and U. S. Navy standard Lighting-off and Securing Sheets (*See* Appendix B).

Inspections

1. *Weekly*
    a. Run turbines with steam, if available, or turn idle turbines and circulate oil by hand.
    b. Lubricate trip mechanism, regulating valve, etc., as applicable.
    c. Check condition of lubricating oil.
    d. Operate relief valves by hand.
    e. Test overspeed trip or speed-limiting governors, as applicable. This should also be done each time a turbine is put in operation after having been secured.
2. *Quarterly*
    a. Test relief valves, if installed.
    b. Check thrust position of turbine rotor.
3. *Annually*
    a. Examine interior through inspection ports, when provided.
    b. Examine for loose or broken bolts. Inspect turbine casing exterior for corrosion.
    c. Clean steam strainers, if fitted.
    d. Check condition and clearances of thrust and sleeve bearings, where accessible.
4. *During Shipyard Overhaul*
    a. Check gland packing for wear.
    b. Examine ball bearings, where fitted. Check condition and clearance of all thrust and sleeve bearings.
    c. Steam-test relief valves.

Note: These tests and inspections are considered the minimum requirements to give adequate assurance of safe and reliable operation of the turbo-generator. Any sign of improper operation of equipment should indicate the need for a test to determine the cause of irregularity. Additional or more frequent tests may be performed at the discretion of the engineer officer.

Safety Precautions

1. *For turbo-generators*
   a. Turn turbine by hand daily and before admitting steam to the turbine casing.
   b. Do not lash down an overspeed trip or a speed-limiting governor, or take other steps to render them inoperative.
   c. Keep the exhaust-casing relief valve set at the proper pressure and in operating condition at all times.
   d. Keep the oil reservoir filled with clean oil at all times.
   e. Before starting the turbine inspect to see that it is clear of foreign matter, especially if the unit has not been operated for a long period.
   f. Keep the turbine unit properly balanced at all times.
   g. Avoid water hammer by properly draining lines and opening valves slowly.
   h. Before turbine is put in service, test the overspeed trip, if provided.
   i. Test the speed-limiting governor by steam at least once a quarter.
   j. Avoid passage of steam through a turbine with the rotor at rest.
   k. Avoid air being drawn through generator

turbine glands with the rotor at rest.

l. Keep the governor-operating mechanism and regulating-valve stems clean and free from corrosion, and the cylinder insulation well clear of the steam-chest lift rods.

m. Maintain designed bearing oil pressure.

n. Keep oil strainers clean.

o. Keep the oil cooler clean.

p. Be sure that steam lines are properly drained. Numerous turbo-generators have been damaged in the past due to the carry-over of slugs of water in the steam when starting up generators in emergencies before steam lines had been properly drained.

2. *For switchboards*

a. Always be sure that protective grab rods and guard rails around switchboards and other equipment are in position when the equipment is energized, unless emergency repairs are necessary while the equipment is in service. Grab rods and guard rails should be carefully maintained to ensure that they are secure and will not accidentally be dislodged. The insulated floor mattings provided as deck covering in the front and rear of switchboards should always be in place.

b. Be careful, when maintenance work is performed on an electric circuit, to ensure that the circuit remains dead and is not energized by the closing of a remote circuit breaker. All circuit breakers or switches which would energize the circuit, if closed, should be marked with a warning sign (*see* NavShips *Technical Manual,* Chapter 9600).

c. Exercise great caution in removing covers (plates or grillages) on switchgear units, when the equipment is energized. For example, unless care is exercised, a large cover on the rear of a switchboard section may tip in at the top or bottom after removal of some of the fastenings or while it is being lifted off. This may result in contact with a live circuit, thereby causing a short circuit, an arc, or injury to personnel.

Maintenance and Operation

1. *Overspeed trips*

To ensure that turbine-driven generators will not run at speeds greater than that for which they are designed, turbines are equipped with a constant speed (speed-regulating) governor and an overspeed trip which shut off steam to the unit after a predetermined speed has been reached. This predetermined speed is about 110 percent of normal operating speed.

2. *Casualties*

a. Turbine end

Casualties to the turbine end are the same as those detailed on page 35 for main turbines, less the lock shaft procedure.

b. Generator end

NavShips *Technical Manual,* Chapter 9610, contains a complete cause-and-correction trouble-shooting guide for such problems as:

(1) Vibration
(2) Hot bearings
(3) Oil leaks
(4) Noisy brushes
(5) Scoring
(6) Blackening of commutators and collecting rings
(7) Failure to generate full voltage
(8) Failure to generate
(9) Open circuits

3. *Characteristics of electrical installations*

The characteristics built into naval electrical installations are simplicity, ruggedness, reliability, and flexibility. These permit continued service after a part of the equipment has been damaged. It is the function of those who operate these plants to make full use of these inherent capabilities, and to maintain, as far as possible, uninterrupted availability of electric power.

4. *Switchboards*

a. Electric power is used on modern naval ships to furnish numerous services that are indispensable to its fighting or functional effectiveness. A ship with all electrical power lost is nearly devoid of value as a fighting unit.
b. The distribution system is the vital link connecting the generators that furnish electric power to the equipment that uses it. Its purpose is to transmit power from one place to

another and, by means of circuit breakers and fuses, to protect the generators from damage. It also affords some protection to the equipment that uses electric power. The main components of a distribution system are:

(1) Ship service switchboards
(2) Bus-ties
(3) Power distribution cables
(4) Power supply and controls for main ventilation fans
(5) Fine control of ship service and emergency power for both damage control and maintenance.

c. The emergency power system is designed to furnish an automatic source of emergency power when the main power distribution system fails. It is equipped with

(1) Interconnections between the emergency and ship service distribution systems
(2) Emergency switchboards

d. The casualty power system is designed to provide temporary connections and cabling to span damaged portions of the ship.

5. *Preventive maintenance*

a. Three fundamental rules for the maintenance of electrical equipment are:

(1) Keep equipment clean and dry.
(2) Keep electrical connections and mechanical fastenings tight.
(3) Inspect and test at sufficiently short intervals to make sure that the equipment is in operating condition.

b. The equipment in the distribution system which requires maintenance can be grouped in two general classes: cables with their fittings and switchboards (including distribution panels, etc.) with their associated equipment. For instructions on the maintenance of both classes of equipment, *see* NavShips *Technical Manual,* Chapter 9600.

6. *Instructions for specific items*

Instructions for specific items of electrical equipment can be found in NavShips *Technical Manual* as follows:

Tables of Engineering Data	Chapter 9005
Electric Propulsion Installations	Chapter 9410 Section IV
Electric Plant-General	Chapter 9600
Electric Generators and Voltage Regulators	Chapter 9610
Distributions Systems	Chapter 9620 Section I
Portable Storage Batteries and Dry Batteries	Chapter 9620 Section II
Electric Motors and Controllers	Chapter 9630
Lighting	Chapter 9640
Interior Communication Installations	Chapter 9650
Search Lights	Chapter 9660
Electronics Equipment	Chapter 9670
Electrical Measuring Instruments	Chapter 9690
Fire-Control Installations	Chapter 9710
Degaussing Installations	Chapter 9810 Section III
Welding and Allied Processes	Chapter 9920 Section III

SUMMARY

In this chapter, an attempt has been made to classify the major equipment found in a steam-power propulsion plant. Needless to say, three volumes of NavShips *Technical Manual* and several books have been written on this subject and the authors of this Guide have attempted only to review some of the major points. An engineer has to work at his trade until operating and maintenance procedures and casualty symptoms become second nature to him.

In addition, the engineer must familiarize himself with various shipboard hull auxiliary equipment and other engineering subsystems described in NavShips *Technical Manual:*

Winches and Capstans	Chapter 9200
Steering Systems	Chapter 9220
Commissary Equipment	Chapter 9340
Laundry Equipment	Chapter 9350
Shafting and Bearings	Chapter 9430
Lubricating Systems	Chapter 9450
Pumps (General)	Chapter 9470
Compressed Air Plants	Chapter 9490
Auxiliary Steam Turbines	Chapter 9500
Clutches and Couplings	Chapter 9570
Distilling Plants	Chapter 9580
Refrigeration Plants	Chapter 9590
Electric Motors and Controllers	Chapter 9630

Familiarization can be achieved by cross-referencing NavShips *Technical Manual* with manufacturers' instruction booklets and various NavPers publications, Ships General Information Booklet, and Damage Control Booklet.

CHAPTER 3

Engineering Department Organization and Administration

The organization of a ship's Engineering Department is designed to ensure the operational readiness of the main propulsion plant, machinery auxiliary equipment, and hull auxiliaries with related functions. The organization shown in Figure 6 is typical of the Engineering Department in a large combatant ship. Most ships smaller than a guided-missile frigate (DLG), with low officer complement, combine two or more functions. For instance, in a destroyer type, the main propulsion assistant is also the B Division and M Division officer, and he has no assistants. His chain of command goes directly down to the B Division and M Division leading petty officers.

Since the organization varies for each ship type, this section will provide data for a large combatant organization, but it should be understood that smaller ships have assigned dual responsibilities to personnel within the Engineering Department.

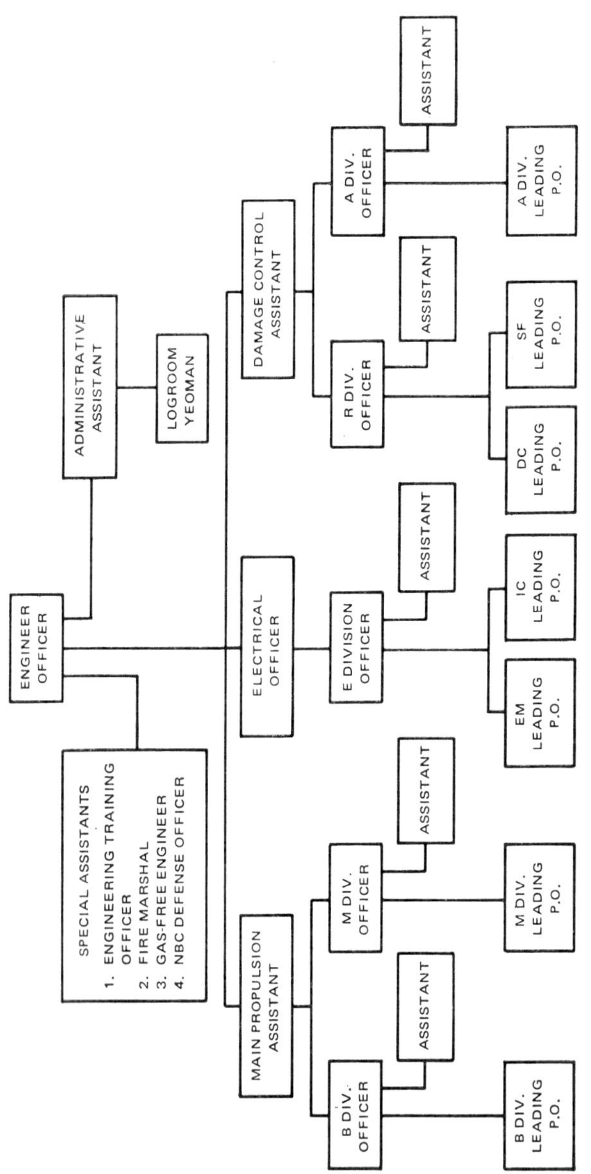

NavPers 10788A, *Principles of Naval Engineering*

Figure 6—Typical Engineering Department Organization For a Large Combatant Ship

To attain maximum efficiency within the Engineering Department, it is necessary that all personnel clearly understand the organizational relationships and the duties, responsibilities, and authority of department personnel.

ENGINEERING DIVISIONS

The major organization unit of a shipboard department is the division. It is at this level that work assignments, administrative procedures, and daily routines are handled. Division responsibility can be given to a newly commissioned officer and, in some cases, to a rated petty officer, depending on the size of the ship. For example, many minesweepers (MSC/MSO) have rated petty officers in charge of divisions. Figure 7 shows the basic administrative organization of a division.

Divisions are normally responsible for the cleanliness, preventive and corrective maintenance, and operation of assigned equipment, and for the cleanliness of the compartment or vicinity in which the equipment is located as follows:

1. *A Division*
   a. air-conditioning machinery
   b. air compressors
   c. anchor windlass
   d. internal-combustion and diesel engines
   e. refrigeration machinery
   f. steering gear
   g. winches and cranes (maintenance only)
   h. laundry and dry-cleaning equipment (maintenance only)

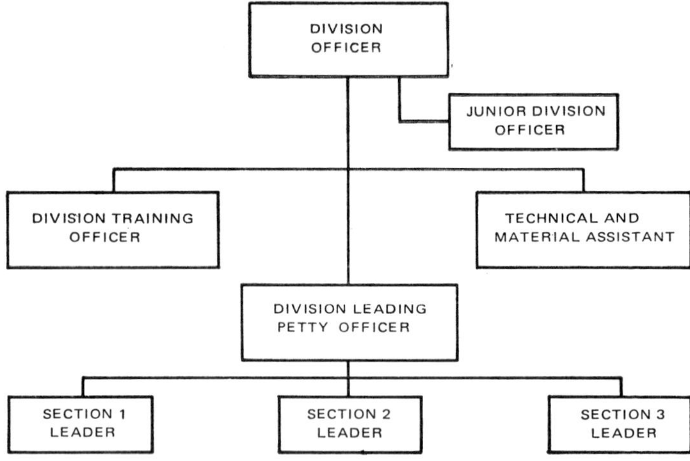

NavPers 10788A, *Principles of Naval Engineering*

Figure 7—Basic Organization of a Shipboard Division

i. galley machinery (maintenance only)
j. emergency fire pumps (maintenance only)
k. boat engines
l. distilling-plant equipment
m. auxiliary boilers
n. hydraulic systems
o. ventilation and heating systems

2. *B Division*
    a. boilers
    b. fireroom auxiliaries
    c. forced-draft blowers
    d. fuel and water testing equipment
    e. fueling-at-sea equipment from the deck connection inboard
    f. fuel-oil piping and pipe fittings
3. *E Division*
    a. all electric motors

b. generators (electric end)
c. controllers
d. electrical distribution systems, including cabling, switching, and protective equipment (no compartment responsibility)
e. gyrocompass and related equipment
f. battery-charging equipment
g. small boat electrical system
h. interior communications systems
i. underwater log system
j. motion-picture equipment
k. television/telephone systems, if installed (repair only)
l. ship-control and indicating systems (maintenance only)
m. portable announcing system (maintenance only)
n. portable electric tools (maintenance only)

4. *M Division*

a. main engines
b. main propulsion equipment, including shafting and engineroom auxiliaries
c. ship service generator units (prime mover side)
d. engineroom valves and piping

5. *R Division*

a. shop-support equipment, including welding and joining
b. damage-control equipment
c. fire-fighting equipment
d. hull fittings
e. piping systems not otherwise assigned

ENGINEERING PERSONNEL

The number of officers and enlisted men assigned to the Engineering Department is determined by the manning requirements for the ship class, as set forth by the Bureau of Naval Personnel.

ENGINEER OFFICER (Engr. Off.)

General Duties

The head of the engineering department of a ship shall be designated the engineer officer. In addition to those duties prescribed elsewhere in these regulations for the head of a department, he shall be responsible, under the commanding officer, for the operation, care and maintenance of all propulsion and auxiliary machinery, the control of damage, and upon request of the head of department concerned, the accomplishment of those repairs which are beyond the capacity of the repair personnel or equipment of other departments but within the capacity of the engineering department. [*Naval Regulations,* 1948, Article 0946.]

Specific Duties

The engineer officer, under the commanding officer, shall be responsible for the proper performances of the functions of his department, which include:

1. The operation, care, and maintenance of all machinery, piping systems, and electric and electronic devices not specifically assigned to other departments.
2. Damage control.
3. The repair of the hull and its appurtenances.
4. The furnishing of power, light, ventilation, heat, refrigeration, compressed air, and water; and the operation, care and maintenance of the equipment connected therewith.
5. The operation, care and maintenance of boat machinery.
6. The care, stowage, and use of fuels and lubricants not assigned to other departments.
7. The maintenance of underwater fittings.

8. The maintenance of the engineering log and the engineer's bell book. [*Navy Regulations,* 1948, Article 0947.]

In other words, the engineer officer represents the commanding officer in all matters pertaining to the Engineering Department. His orders to the personnel in his department must be obeyed. Under the commanding officer, he is specifically responsible for the operation, care, and maintenance of the ship's

propulsion plant
electric power plant
auxiliary machinery
piping systems
interior communications systems
damage-control posture
hull repairs
material and equipment beyond the capacity of personnel in other departments when requested by the cognizant department head.

In the performance of his duties, the engineer officer must:

1. conform to policies and comply with orders set forth by the commanding officer and/or the executive officer.
2. use his privilege of conferring directly with the commanding officer on matters relating to engineering operations, if he believes such action necessary for the good of the department. He must, however, keep the executive officer informed on such matters.
3. normally report to the executive officer for matters relating to administration of the Engineering Department.

4. keep the commanding officer informed of the operational readiness and actual operation of the propulsion and electrical plants and of the damage-control systems.
5. obtain permission of the commanding officer prior to placing out of commission any equipment without which the ship's safety or operational readiness could be adversely affected.
6. perform general duties which include:
   a. proper administrative organization of the Engineering Department
   b. training and assignment of personnel
   c. frequent inspections of department personnel, material, and spaces to ensure readiness
   d. preparation and updating of various Engineering Department instructions, bills, and safety precautions
   e. control funds allotted to the Engineering Department and ensure that they are spent to best advantage
   f. cooperation with the heads of other departments in order to contribute to the coordinated command effort
   g. proper documentation of Engineering Logs.

MAIN PROPULSION ASSISTANT (MPA)

General Duties

The main propulsion assistant, when assigned, shall be responsible under the engineer officer for the operation, care, and maintenance of the ship's propulsion machinery, auxiliaries related thereto, and such other auxiliaries as may be assigned. [*Navy Regulations,* 1948, Article 0951.]

Specific Duties

> The main propulsion assistant, under the engineer officer, shall be responsible for:
>
> 1. The effective operation of the main engines and boilers and assigned auxiliaries. The main propulsion assistant shall make, or cause to be made, frequent examinations of the machinery and equipment for which he is responsible, and shall insure that necessary repairs and adjustments are effected, subject to such authorization as may be required. On getting underway, coming to anchor, and at other times when unusual care is required, he shall personally supervise the operation of the main engines. He shall insure that fires are not lighted or secured under a boiler except in cases of emergency, without permission of the commanding officer. He shall insure that the main engines are not turned except in obedience to a signal from, or by permission of, the officer of the deck.
> 2. The care, stowage, and use of bunker fuels, except those for aircraft, and the operation, maintenance, and security of the systems pertaining thereto; the keeping of fuel records and the submission daily to the commanding officer of a fuel report.
> 3. The preparation and care of the engineering log and the engineer's bell book. He shall daily, and more often when necessary, inspect them and shall take such corrective action as may be necessary, and within his authority, to insure that they are properly kept.
> 4. The preparation of such operation, maintenance, and other engineering records as may be necessary or as may be prescribed. [*Navy Regulations,* 1948, Article 0952.]

The MPA is responsible, under the engineer officer, for the operation, care, and maintenance of the ships' main propulsion machinery and assigned auxiliaries. In the performance of his duties, he must

1. learn the operating characteristics of equipment under his cognizance.

2. make frequent inspections of all machinery and equipment under his cognizance.
3. attempt to be at the main control station during special sea details and general quarters exercises, in order to supervise the machinery for which he is responsible.
4. assume the responsibilities of a division officer when B Division and M Division officers are not assigned.
5. assume responsibility for
   a. care, stowage, and expenditure of fuel
   b. proper preparation, care, submission, and disposition of the Engineering Log, the Engineer's Bell Book, and other engineering records assigned to his care.

DAMAGE CONTROL ASSISTANT (DCA)

General Duties

The damage control assistant of a ship, when assigned, shall be responsible, under the engineer officer, for establishing and maintaining an effective damage control organization, and for supervising repairs to the hull and machinery, except as specifically assigned to another department or division. [*Navy Regulations,* 1948, Article 0955.]

Specific Duties

The damage control assistant, under the engineer officer, shall be responsible for:

1. The prevention and control of damage, including control of stability, list, and trim. He shall supervise placing the ship in the condition of closure ordered by the commanding officer. He shall insure that appropriate closure classifications are assigned and conspicuously marked upon or adjacent to the objects to which they apply. He shall coordinate and supervise the carrying out of prescribed tests of com-

> partments and spaces for tightness. He shall prepare and maintain bills for the control of damage and stability, and shall insure that correct compartment check-off lists are posted.
>
> 2. The training of ship's personnel in damage control including fire fighting, emergency repairs, and nonmedical defensive measures against gas and similar weapons.
>
> 3. The operation, care, and maintenance of auxiliary machinery piping, and drainage systems not assigned to other departments or divisions, and of the shop repair facilities; and the repair of the hull and boats. [*Navy Regulations,* 1948, Article 9056.]

The DCA is responsible, under the engineer officer, for the establishment and maintenance of an effective damage-control organization and for the supervision of repairs to the ship's hull, machinery, and piping systems, as assigned. In the performance of his duties, he must

1. prepare directives for any damage-control functions that require interdepartmental coordination.
2. specify the requirements for "all hands" damage-control training, and prepare the ship training syllabus.
3. inspect damage-control equipment and ensure its proper maintenance.
4. inspect watertight integrity throughout the ship and ensure its maintenance.
5. be sure that compartment check-off lists are current.
6. organize and train damage-control parties to meet all anticipated emergencies.
7. inform the engineer officer of shipboard practices that lower damage-control readiness.
8. make himself familiar with the general ar-

rangement of the ship so that he can effectively control ship damage while in charge of Damage Control Central.

9. make himself familiar with the NBC warfare passive-countermeasures procedures.
10. assume duties of A, E, and R division officers when such are not assigned.

ELECTRICAL OFFICER (ELEC. OFF.)

General Duties

> Assistants to the engineer officer may include . . . an electrical officer, who shall be responsible under the engineer officer for the operation, maintenance, and repair of the electrical machinery and systems throughout the ship, except as assigned to another department. [*Navy Regulations,* 1948, Article 0948.]

The electrical officer is responsible, under the engineer officer, for the operation, care, and maintenance of the ship's electrical power generators and distribution systems, interior communications systems, degaussing equipment, gyrocompass and associated equipment, any small-boat electrical systems, and entertainment equipment. He is also responsible for

1. the preparation and submission of the logs, records, and reports required in connection with his assigned duties
2. collateral duty of motion-picture officer.

DIVISION OFFICER (DIV. OFF.)

Definition

> A division officer, within the meaning of these regulations, is one regularly assigned by the commanding officer to command a division of the ship's organization. [*Navy Regulations,* 1948, Article 1043.]

Responsibilities and Duties

1. A division officer shall be responsible, under the head of his department, for the proper performance of the duties assigned to his division, and for the conduct and appearance of his subordinates, in accordance with regulations and the orders of the commanding officer and other superiors. He shall keep himself informed of the capabilities and needs of each of his subordinates and, within his authority, he shall take such action as may be necessary for the efficiency of his division and the welfare and morale of his subordinates. He shall train his subordinates in their own duties and in the duties to which they may succeed, and shall encourage them to qualify for advancement and to improve their education. He shall suppress any improper language or unseemly noise or disturbance, and he shall report to the executive officer all infractions of regulations, orders, and instructions which are deserving of disciplinary action.
2. He shall by personal supervision and frequent inspection insure that the spaces, equipment, and supplies assigned to his division are maintained in a satisfactory state of cleanliness and preservation. He shall report promptly to his head of department any repairs which may be required or other defects which need correction and which he is unable to effect.
3. He shall carefully instruct his subordinates in all applicable safety precautions and shall require their strict observance.
4. He shall maintain a corrected copy of the watch, quarter, and station bill and other bills and orders for his division and shall insure that pertinent parts thereof are kept posted where they will be accessible to his subordinates. [*Navy Regulations,* 1948, Article 1044.]

Each division of the Engineering Department is commanded by an officer or senior petty officer appointed by the commanding officer upon the recommendation of the engineer officer. It is the responsibility of all the division officers (A, B, E, M, and R) to assist the DCA and MPA in the performance of their duties. Specific duties, responsibilities, and authority of each division officer of the Engineering Department include:

1. directing the division operations through leading petty officers.
2. assigning division personnel to watch stations and duties.
3. scheduling of division training.
4. evaluating enlisted personnel performance.
5. maintaining personnel records.
6. ensuring proper preparation, maintenance, and submission of cognizant reports and correspondence.
7. ensuring that security and safety precautions are strictly observed.
8. recommending personnel for transfer and for advancement in rate.
9. handling of leave, liberty, and special privileges.
10. conducting periodic inspections for the purpose of evaluating the performance of enlisted personnel.

When large ships are over their manpower allowance, a few inexperienced junior officers may be assigned as Junior Division Officers or Division Training Officers. This is not a common occurrence in the fleet today.

MISCELLANEOUS OFFICER ASSIGNMENTS

Many miscellaneous shipboard duties are assumed as collateral assignments by either an officer or a qualified petty officer as follows:

Engineering Administrative Assistant

He is responsible for

1. operation of the log room
2. assignment, training, and evaluation of log-room yeoman
3. maintenance of the Engineering Department's records
4. maintenance of the department's "tickler" file on required reports
5. preparation of the department's watch bills
6. screening incoming correspondence and initiating action as appropriate
7. assisting in preparation of Engineering Department directives
8. keeping log-room files, such as blue prints, instruction booklets, NavShips *Technical Manual,* and other publications up to date.

Engineering Training Officer

He is responsible for

1. assisting the engineer officer to develop a departmental training program to support the training objectives of the ship
2. coordinating division training programs
3. implementing ship training within the Engineering Department
4. keeping departmental training records and reports
5. assisting in the training of junior officers and key petty officers of the Engineering Department.

NBC Defense Officer

He is responsible for

1. providing technical advice in relation to NBC defense
2. assisting division officers in training personnel in NBC defense
3. maintaining effective countermeasures for combating NBC attack.

Gas-Free Engineer

He is responsible for

1. analyzing conditions in the ship to ensure against danger from suffocation or hazard from explosive gases
2. organizing and administrating a training program for informing all hands in shipboard hazards
3. advising the engineer officer and commanding officer of any hazards relating to specific welding operations.

Fire Marshal

He is responsible for

1. elimination of fire hazards aboard ship
2. maintenance and reliability of appropriate fire-fighting equipment in the ship
3. assisting the DCA in training ship personnel in fire-fighting procedures.

ENLISTED PERSONNEL

The enlisted complement and allowance of the Engineering Department consists of many technical specialists. For convenience, the ratings are listed below, but NavPers training books give detailed descriptions of the specialties involved. The senior ratings are the real backbone of the Engineering Department. The intimate knowledge and experience of these dedicated enlisted personnel are valuable sources of information for the less experienced personnel, and the engineer officer could not function efficiently without them.

Boilermaker	BR
Boiler Technician	BT
Electrician's Mate	EM
Engineman	EN
Hull Maintenance Technician	HT
Interior Communications Electrician	IC
Machinery Repairman	MR
Machinist's Mate	MM
Molder	ML
Patternmaker	PM
Fireman	FN
Yeoman	YN

ENGINEERING RECORDS AND REPORTS

Records maintained by the Engineering Department and reports submitted to the engineer officer provide data for engineering reports to higher authority. Comparison of past and present reports, records, and logs will provide a knowledgeable officer with a clear picture of the condition of the engineering plant. It will often even reveal pending maintenance requirements prior to the occurrence of a major casualty.

Since the evaluation of this information is based upon records, the evaluation is only as complete as the records themselves and is dependent on the integrity of the personnel who prepared them. Therefore, the maintenance of accurate, legible, and current engineering records and reports is essential and reflects efficient administration of the Engineering Department.

There is no simple way to ensure accurate reports. Once responsibility for record keeping is fixed, only concerted training of personnel, coupled with appropriate reward or punishment, will ensure the integrity of the reports and logs. Proper training to correctly obtain, interpret, and record data will ensure accuracy. The two required engineering legal records are discussed in detail in Chapter 6.

Engineering operating records are designed to ensure regular inspection of operating machinery and to provide data for performance analysis. They are not a replacement for periodic visual inspections by leading petty officers or division officers. Type commanders' directives specify the details and requirements for engineering logs, and books such as NavPers 10788A provide sample logs and describe entries. This Guide is not intended to provide such detail. Instead, it lists the most common operating logs which are retained on board for a period of two years, then catalogued and stowed for ready access when required.

RECORD	FORM SOURCE	CONTENT
Propulsion Steam Turbine Operating Record	NavShips 3652	A daily record maintained for each propulsion steam turbine in operation.

RECORD	FORM SOURCE	CONTENT
Diesel Engine Operating Record	NavShips 3648	A daily record maintained for each operating auxiliary or propulsion diesel engine.
Boiler Room Operating Record	NavShips 3651	A complete daily record maintained for each steaming fireroom.
AC/DC Electro Propulsion Operating Record	NavShips 3647	A daily record maintained for each operating propulsion generator and motor equipped with AC or DC electro propulsion machinery.
Electrical Log	NavShips 3649	A complete daily record maintained for each operating ship's service and emergency generator. Contains data on both prime mover and generator end.
Distilling Plant Operating Record	NavShips 3676, 3676-1, or 3675	A complete daily record maintained for the installed distilling plant.
Refrigeration/Air Conditioning Equipment Operating Record	NavShips 4731	A complete daily record maintained for each operating refrigeration and air-conditioning plant.
Air Compressor Operating Record	Tycom Form	Contains important data such as temperature and pressures pertaining to air compressors in operation.

RECORD	FORM SOURCE	CONTENT
Gyrocompass Operating Record	Tycom Form	Complete daily record for recording times of starting and stopping the gyrocompass. Includes total hours operation.
I.C. Room Operating Record	Tycom Form	A daily record of major electrical equiment in operation in the IC room.
Fuel and Water Report	NavShips 115	Records daily use of fuel oil, lube oil, water, and diesel oil. Provides daily status of these liquid loads.
Engineering Log	NavShips 117	Legal record, discussed in Chapter 6.
Engineer's Bell Book	NavShips 116	Legal record, discussed in Chapter 6.
Fueling Memorandum	Locally prepared	A report to CO to inform him and others aboard of the amount of fuel received or delivered.
Boat Fueling Record	Locally prepared	A daily record of the fueling of power boats.
Oil King's Memorandum	Locally prepared	A report of fuel oil and feedwater status as of 1000 and 2000 hours. Submitted to engineer officer by Oil King.
Monthly Summary of Fuel Inventory and	NavShips 5083	A comprehensive summary of the ship's fuel inventory, fuel con-

RECORD	FORM SOURCE	CONTENT
Steaming Hours Report		sumption, and steaming hours.*
Engineering Officer's Night Order Book	Locally prepared	Contains engineer officer's standing night orders with respect to operation of the engineering plant, special orders or precautions concerning speed and operation of main engines, and all other orders for the night's steaming to the Engineering Officer of the Watch (EOOW).
Steaming Orders	Locally prepared	Lists major machinery units and readiness requirements for Engineering Department in relation to getting underway.
Boiler Record Sheet	NavShips 114	Monthly summary of boiler's operation and includes boiler water tests. Includes hours steaming since last fireside and waterside cleaning.
Main Propulsion Turbine Condition Report	Letter report	A letter on turbine condition submitted to NavShips, via Tycom, three months prior to regular overhaul. Includes recommendations on turbine repairs.

* This report is vital and accuracy is required since it provides current and cumulative data necessary for determining budget and fleet logistic requirements and operational performance.

INSPECTIONS

Ship inspections are conducted to determine readiness for service. Attempts to hide problems from the inspectors can only harm the ship in the final analysis. This philosophy may not suit some but in the authors' opinion, "honesty's the best policy." A satisfactory grade and an unready ship somehow are incompatible. The satisfactory grade should be earned by being well prepared for the inspection.

Readiness inspections are normally conducted by the type commander under a designated subordinate as chief inspector. The organization of the inspection board normally conforms to the administrative organization of the ship. Check lists to assist the inspection party and the ship are normally provided by the type commander. The evaluation of readiness is based upon how well the ship meets the established standards for efficient accomplishment of its mission.

ADMINISTRATIVE INSPECTION (ADMIN)

The administrative inspection is designed to determine how efficiently and effectively prescribed administrative procedures, including current type commander directives, are being implemented. The inspection of the Engineering Department is divided into two parts:

1. General ship evaluation, which includes
   a. appearance and bearing of personnel
   b. cleanliness, sanitation, and appearance of ship
   c. adequacy of personnel's clothing

   d. crew's knowledge of routine procedures and shipboard organization
   e. adequacy of method of information dissemination and crew indoctrination
   f. adequacy of training and educational facilities for individuals.
   g. habitability of crew's quarters.

2. Engineering Department specific evaluation, which includes
   a. departmental logs, bills, files, blueprints, records, and general paper work
   b. cleanliness and preservation of engineering spaces
   c. adequacy of training engineering personnel
   d. posting of safety precautions, warning signs, and guards
   e. labeling of piping systems
   f. proper assignment of watch standers
   g. proper maintenance of operating logs.

MATERIAL INSPECTION (MAT)

The material inspection is designed to evaluate the actual material condition of the ship, including proper functioning of all machinery and fittings. The results of this inspection can disclose those repairs, alterations, or changes which are needed to ensure material readiness and can reveal whether or not proper maintenance procedures have been followed in the care and operation of machinery and equipment.

Basic requirements for material readiness are:

1. established routines for conducting inspections

and tests, planned maintenance procedures, and a system to ensure rapid repair response time

2. planned and effective use of ship's facilities for preservation, maintenance, and repair
3. adequate material maintenance records
4. proper work request preparation and work allocation on equipment requiring repairs (i.e., ship's force, tender, or shipyard work)
5. proper documentation and use of detailed maintenance and repair procedures.

OPERATIONAL READINESS INSPECTION (ORI)

This inspection is conducted to evaluate the offensive and defensive capabilities of a ship at any given moment. It observes and grades an actual battle problem, then determines existing weaknesses and deficiencies, and makes recommendations for improvement. At the conclusion of such an inspection, the Engineering Department's operational readiness is evaluated on three bases:

1. *General Evaluation*

   a. extent of preparation and fulfillment of ordered conditions of readiness as appropriate to drills and battle problem
   b. extent of correct utilization of damage-control features of the ship
   c. extent to which proper engineering casualty control is accomplished
   d. extent to which personnel take appropriate corrective actions for control of damage
   e. adequacy of damage situation reports and dis-

semination of information to other departments, OOD, and Commanding Officer

f. the general handling of the plant in accordance with good engineering practice and its ability to supply all vital services for fighting the ship.

2. *Hit Evaluation*

The evaluators' questionnairc will establish certain criteria which the ship should be prepared to meet in order to ensure a satisfactory evaluation.

a. Was the engineering plant arrangement suitable to meet battle conditions?
b. Was proper condition of readiness set?
c. Were fire and missile hazards present?
d. What was the condition of fire-fighting and damage-control gear?
e. Was personnel clothing and protective clothing worn properly?
f. Were ship personnel in readiness?
g. How prompt and effective was the casualty control procedure?
h. Did personnel use established doctrine and procedures?
i. Were proper reports made?
j. Were alternate engineering systems in readiness?
k. Were safety precautions observed?
l. Were there material deficiencies in equipment?
m. Was there coordination between repair personnel and engineering control spaces?

3. *Main Engine Control Evaluation*
    a. How efficient was receipt of vital communications?
    b. How efficient was origination and transmission of reports to OOD, DCC, and other stations?
    c. Was correct action taken based on sound judgment and good practice?
    d. How fast and efficiently were casualties corrected?

The three inspections described above are considered the major inspections which relate to administration and organization and which fall under the type commander's cognizance. Other inspections for which the Engineering Department should always be ready are "personnel inspection" and "below decks" inspection. The better the department is prepared for ship-type inspections, the easier it will be to meet the type commander's inspection requirements. A well-trained ship does not have to prepare for inspections on a crash basis: it stays ready. The idea is not to put on a show for a day or two, but to truly "be prepared" to meet any and all situations as they arise.

Many other inspections, which can properly be called trials, are discussed in Chapter 10.

PERSONNEL PROBLEMS

Officers and leading petty officers in the Engineering Department must be able to cope with the many special personnel problems that occur aboard ship. This section is developed to point out some of the problem areas, around which many texts on leadership, small-group theory, and psychology have been written. Yet, the problems still exist and no "cure-all" has been found.

The engineer's dilemma is to deal with such problems effectively and in a timely manner without damaging department morale. And, although each individual problem may be small, the cumulative effect can account for much of the division officer's working day. Typical problems are:

1. Many men always have a standby request submitted. Remember that men who are constantly away from their in-port duty stations tend to lose the training alertness required in emergencies. Standbys are a privilege, not a right. The ship must have qualified men aboard in the watch section, but legitimate requests should be considered on their merit.
2. Drinking, carried to the point where a man is insensible or not in complete possession of his faculties, is a recurring Navy problem. Intoxication in a foreign port is a particularly serious problem because such behavior by men in uniform brings discredit on our country. The OOD must be alerted to inform the duty engineer of any engineering personnel who attempt to enter engineering machinery spaces on returning from liberty in this condition. Drunken carelessness in getting a plant ready for underway did, in fact, cause the destruction of an entire boiler on a fleet destroyer in the 1950s.
3. The engineer officer must be alert to agreements made among the crew (for example, "standby for me on the request, but I'll be back for my

watch"). Either the standby takes the full duty or the request should be denied.

4. Many naval men marry young. After leaving on an extended cruise, it is not uncommon for the ship to start receiving indebtedness letters from businesses in the home port. Sometimes this is due to the man's own poor management, and sometimes to the inexperience of his wife who has been left alone and is unable to cope with the business affairs. The best one can do is counsel the man in trouble, advise him whether he is liable to be charged under the Uniform Code of Military Justice (UCMJ), and try to get him to handle his affairs properly. Recommending that his wife go to Navy Relief and/or the base chaplain is always worth a try.
5. Many moral problems are encountered in the shipboard environment. Leading personnel of the Engineering Department should set a good example for the men. Practice what you preach. Don't be a hypocrite.
6. Drug usage is a major problem. Locate offenders and pursue the problem as detailed in current Navy Regulations and directives from higher authority.
7. Divorces or initiation of divorce proceedings occur as frequently in the service as in civilian life. The best advice is to recommend to the persons involved that they obtain marriage counseling. Do not become personally involved.
8. Requests for emergency leave should be granted expeditiously when warranted.

These are but a few of the personal problems encountered in the Engineering Department. The division officer and leading petty officers must try to help their men through crises, but under proper Navy procedures. Violations of authority by senior officers only set poor examples. There is no set way to deal with personal problems, but a few possible techniques are discussed in Chapter 4.

SUMMARY

The information provided in this chapter relating to administrative, organizational, and functional relationships within the Engineering Department should be of value to the junior officer or leading petty officer who finds himself thrust into a position of leadership and responsibility. In addition, the duties of key engineering personnel, the requirements for recording and reporting engineering data, the various key inspections to be encountered, and some of the division-level personnel problems have been discussed in an effort to aid the inexperienced junior officer in his preparation for shipboard duty.

CHAPTER 4

Management

According to Henri Fayol, a management theorist, the principles of management are planning, organization, direction, coordination, and control. Others have defined management within the acronym POSDCORB which stands for planning, organizing, staffing, directing, controlling, observing, reporting, and budgeting. In any case, within the shipboard organization, the engineer officer serves in a management position as related to the resources of men, material, and money.

There is no need for this Guide to dwell on the quantitative and qualitative techniques that have been developed for use in the higher echelons of business and government. Aboard ship, interests should relate to proper organizational structure for the Engineering Department, social problems relating to personnel management, and proper planning for the expenditure of existing funds. Funds will always be less than the amount desired for—in some cases considered essential to—the proper operation and maintenance of equipment assigned to the department.

PLANNING

Planning is probably one of the most important aspects of management, yet many people fail to realize what it is. In the military, higher echelons have always had predetermined courses of action, and at lower echelons, the planning process has been in effect for years.

At the shipboard level, there are three characteristics that must exist if something is to be considered a plan: it must involve the future, set forth a course of action, and have some personal or organizational identification.

The purpose of a plan is simply to bring about a certain condition. This condition is either a goal or an objective. Therefore, a goal or objective must be the outcome of implementing a plan, but a goal itself does not necessarily constitute a plan.

A formal procedure for establishing plans has been developed over the years, and it can be broken down into two major phases:

1. *Conception*
   a. Becoming aware of a need for a plan
   b. Providing a precise formal statement of the objective of the plan
   c. Preparing a broad outline of the proposal to prepare the plan
   d. Obtaining approval of a formal proposal to prepare the plan.
2. *Preparation*
   a. Organizing a planning staff and assigning responsibility
   b. Determining outline specifics of the plan

c. Establishing liaison with involved units
d. Obtaining data for the plan to meet scheduled requirements
e. Evaluating collected data
f. Formulating tentative conclusions and/or plans and alternatives
g. Testing the components of the plan
h. Preparing the final plan
i. Testing of the final plan
j. Implementing the plan when approved by higher authority.

ORGANIZING

Organization structure must permit good communications, yet provide the authority for individuals to make simple decisions. Committees can be highly effective on complex decision matters but they are relatively inefficient. On the other hand, authoritative control is highly efficient but quite ineffective in any complex decision-making process.

Basically, the Navy is organized on a scalar, or chain-of-command, principle, which involves the grading of duties in line with degrees of responsibility. Navy structure also includes both the coordinative principle of orderly arrangement of group effort to provide unity of action, and the functional principle of distinguishing between various kinds of duties.

Normally, the shipboard organization is already structured through assignment of specific billets by the Bureau of Personnel, and the engineer officer's job becomes one of planning, coordination, and control. But when organization is lacking or needs revising, a basic approach is to follow the work flow of the department. By separating the primary work from auxiliary functions, the department can be structured to

1. secure adequate attention for all phases of its operation
2. ensure all work is assigned to technically competent personnel
3. relieve senior personnel of subsidiary duties
4. prevent uncontrolled growth of subordinate divisions within the department.

KEY FACTORS IN GROUPING ACTIVITIES

If an organizational problem exists, it can usually be solved by:

1. taking advantage of specialization of personnel
2. facilitating control by independent checking among divisions
3. aiding coordination by grouping together dissimilar activities which require coordination
4. securing adequate attention to all department activities by exercising a proper span of control
5. recognizing such human considerations as availability of personnel, training, tradition, and prevailing attitudes
6. properly delegating authority
7. ensuring policy and procedures are in written format
8. ensuring proper line/staff relations are developed
9. establishing proper lines of communications.

A few of the above factors require some detailed discussion.

Span of Control

This factor focuses attention on the realization that humans have limitations as to available time and energy. The number of subjects to which a man can give his attention is limited. Span of control represents the number of people that one person can effectively supervise directly. To establish span of control, one must consider control at all levels of the Engineering Department by analyzing the

1. time to be devoted to supervision
2. variety and importance of activities to be supervised
3. repetitiveness of activities
4. ability of subordinates
5. staff assistance available.

With this information, the engineer officer can actually tailor the span of control for each major position within the department.

Written Policies and Procedures

A *policy* is a guideline for decision making: it is generally broad and usually sets the limits of action. A *method* is a guideline that is more specific than a policy and attempts to exert some control over operating conditions. A *procedure* is a very detailed, specific guideline and is usually sequential in nature.

Policies and procedures must be in writing. If they are not, problems can arise in holding personnel accountable, and there is no check as to whether policy is being verbally changed at lower levels within the department.

Resistance to Change

In implementing changes, human factors must be taken into account. Behavioral patterns indicate that subordinates generally resist change, particularly if it is forced upon them and the reason for making the change has not been made clear. The major reason for this resistance is that the parts of an organization are in themselves social systems which generate

1. ways of working
2. stable interpersonal relationships among personnel
3. common norms and values
4. techniques for coping with situations in the environment concerned.

Thus, the change effort can often fail simply because of man's basic resistance to any change in his behavioral patterns. Often, this is the result of failure of the person directing the change to sell its objectives to the working-level personnel.

Resisting personnel also develop problems in that they fail to accept the reasons for change; consider the standards that the change is intended to achieve unreasonable, or the work measurements inaccurate; dislike facing the unpleasant fact of change; are influenced by group pressures; and resist any change to established customs. This individual resistance can be developed further into formal and informal group resistance.

How, then, can support for changes in an organization be won? First, it is essential that adequate communications be established and that individuals be made to feel that they have had a part in developing the change: it is difficult to oppose one's own recommendations. The organization should also assure that the change will not affect either the individual's or the group's physical, social, or on-the-job needs.

With the above as background, it can be concluded that the pain of organizational change can best be eased by the participation form of management, which provides the following motivational forces:

a. development of strongly favorable behavior for implementing goals
b. personal responsibility for goals and for work toward their implementation
c. cooperative inter-organization attitude
d. high job satisfaction.

Best of all, the above forces have a tendency to reinforce one another.

In addition, participation develops a communications process which is aimed at achieving the organization's objectives. Communications originate at all levels of the organization and are directed upward because the subordinates so desire. Thus, there are virtually no forces to distort information, no need for supplementary communications systems, and good relations are maintained between subordinates and seniors.

The participation approach provides for decision-making throughout the organization and for a proper and efficient upward flow of complete and accurate information. It encourages teamwork in the effort to achieve accepted goals, and tends to develop emotionally and socially mature persons capable of effective interaction, initiative, and leadership.

STAFFING

In the department organization, staffing simply relates to the proper utilization of personnel to accomplish specific tasks. Therefore, staffing can simply be defined as the process by which seniors select, train, and promote subordinates.

The major problem in staffing is the delegation of authority. Delegation provides and assigns a set of duties and gives the necessary authority to carry them out. Along with the given authority, the person to whom the delegated duties are given accepts an obligation or accountability. Remember, no one can delegate his responsibility, only his authority and his duties.

Delegation is one of the pitfalls encountered by the young naval officer. Most shy away from delegation and tend to overload themselves with work. Two major errors that must be avoided are failure to make clear the limits of delegated authority and delegation of authority beyond legal limits. Obstacles to actual delegation lie in the facts that seniors feel they can do the job better themselves, or they lack the ability to direct, or they do not have confidence in their subordinates, or they have a temperamental aversion to taking a chance. At times, subordinates fail to accept delegation for one or more of several reasons: it is easier to ask the boss; they fear criticism; they do not have the information or resources to do the job; they lack self-confidence; and/or they are given inadequate positive incentive.

In spite of the problems, delegation of authority is an essential ingredient for proper functioning of an organization. It is a must for senior shipboard personnel.

DIRECTING

Directing can be defined as the process by which the actual performance of subordinates is guided toward fulfillment of the plan or course of action desired. Thus, directing is the heart of the managerial process, because it is involved with initiating action. Many persons convey the same idea by the words *leading, supervising, ordering,* or *guiding.*

Directing is related to the manner in which a senior influences the actions of his subordinates. First and foremost, it includes the issuing of clear, complete, and concise orders which are within the subordinates' capabilities to accomplish. Second, it implies a continual training activity. Third, it involves proper motivation of subordinates. And, finally, it consists of maintaining discipline and rewarding those who perform properly.

The manner by which a senior directs activities depends upon his own personal traits and the working situation. In some cases, where subordinates are unskilled and need detailed instructions, he may find the direct, simple order advisable. If his subordinates are highly educated, such as may be found in a research activity, he may find the consultative approach is advisable. In cases of emergency, he may assume a "take charge" role and give short, clear, authoritative commands.

Much research has been conducted to determine what type of leader is the most effective. Most leaders fall into one of four categories:

1. The dictatorial leader who accomplishes his duties by instilling into his subordinates a fear of penalties. He maintains a highly critical and negative attitude in his relations with subordinates. He expects good performance. It has been found that although this approach may be good in "short run" situations, it does not provide a solid foundation for continuing performance. It does not provide satisfaction for those being led.
2. The benevolent-autocratic leader who assumes a paternalistic role and, by generating respect and allegiance, forces subordinates to rely on him for their satisfaction. This type of leadership does not develop leadership qualities in subordinates: it results in total dependence on the leader's presence and deterioration of work in his absence.
3. The democratic leader who does not depend on his own capabilities alone but consults with his subordinates. He encourages subordinates to participate in the management processes and to venture forth on their own initiative. This type of leadership results in a cooperative spirit and the development of managerial capabilities in the subordinates.
4. The laissez-faire leader who depends completely on his subordinates. This leader can find his subordinates taking charge and heading in undesired directions.

The traits and skills required for leadership and the approaches to it must be considered in the directing process.

Supervision is another important function of directing. Literally, supervision means *overseeing* and thus implies face-to-face encounters with subordinates. In a simple line organization such as that of a small ship, the men at the department head or division officer level must perform all the functions of management, with the exception that skilled petty officers take on the supervisory function. The supervisor's daily contact with both the working level and his seniors places him in a strategic position. He represents an important medium through which subordinates can communicate with their seniors.

If we consider the term *directing* as previously defined, then leadership is directly related to motivation. Motivation is a very real driving force. In today's society, man's safety needs are basically fulfilled. Thus, other needs are predominant. Motivation, potential for development, capacity for responsibility, and readiness to work toward goals are all inherent in people. A manager's responsibility is to recognize a person's abilities and use them for the good of the organization.

The motivation aspect can be viewed as follows: an individual strives to satisfy his physical needs and requires recognition of his abilities by his seniors and friends. Such recogniton contributes to his status and social acceptance, and to his self-esteem (ego). Everyone needs opportunity for freedom of discretion and creativity, and every individual reacts in some manner to both reward and punishment.

CONTROLLING

Control is the process that measures performance and guides it toward some goal. In basic management theory, this term relates to the duties of a comptroller, whose role is to provide quantitative information for decision-making and to measure the performance of an organization. In a ship organization, a department head can consider himself the comptroller of his department. Therefore, the engineer officer is the comptroller for his department. His mission is to establish, coordinate, and maintain an integrated plan for control of organization operations and to provide cost estimates and expense budgets. In formal business organizations the role of the comptroller is greatly expanded, but the above-stated functions suffice for a shipboard organization.

The engineer officer can control effectively if his departmental organization possesses

1. a resource-allocation system
2. a method-of-work authorization system
3. a cost-collection structure or method of financial recording
4. a means of measuring work progress
5. a technique for arriving at budget estimates.

The systems mentioned above must interact so that they implement each other.

ROLE OF THE COMPTROLLER IN PLANNING

In addition, the engineer officer has a responsibility to see that every plan of his department is sound and that all parts of each plan fit together. He must judge any plan on an overall basis:

1. Is it realistic, based on past experience?
2. Does it reflect economic conditions expected?
3. Does it adhere to management policy?

DEVELOPMENT OF THE CONTROL SYSTEM

In developing control, the engineer officer or his designated representative must wisely determine major areas of concern or "strategic control points." To attempt to control everything tends to increase unnecessary efforts and to decrease attention to important problems.

In addition, the need to adjust future actions requires that information based on past performance be obtained. Such information is called *feedback.*

The feedback system and other controls must be tailored to fit the organization. In some cases, units may be planned to control themselves and yet be tied into an overall departmental control system so that direct control can be exerted when necessary.

In all cases, human factors must be considered. A technically well-designed control system may fail because subordinates react unfavorably to it. The subject of human factors as they relate to organizational management are discussed at the end of this chapter.

OBSERVING

The observing function is actually tied directly to the controlling function in that direct supervision is one of the tools of controlling and directing. This should not be a surprise to the reader, for by now, it should be apparent that the functions of management (POSDCORB), as previously defined, form an interrelated system. When observing, the surveillance should be as unobtrusive as possible. It should be done in a manner that does not offend the subordinates. A properly planned, friendly tour through the machinery spaces can provide much observed data.

REPORTING

Reporting is really a simple word meaning that an organization must have communications. It is the process by which ideas are transformed into orders and by which compliance with these orders is assured.

Communication is essential to the functioning of any organization. It is viewed widely as one of the more important management functions. Communication is in itself a complex process which is further complicated by

1. the belief that all material transmitted is understood and accepted
2. the diverse nature of material transmitted
3. unfavorable attitudes by persons in the organization. Hostility, fear, distrust, and similar attitudes tend to reduce the flow of relevant information.

In view of the complex nature and varied character of material that is transmitted, it is not surprising that the communication process often breaks down. Furthermore, distrust and lack of confidence lead some members at all levels in a departmental organization to "play it close to the chest," to share a minimum of information with others, and to view information from other sources with suspicion. Such distrust leads to communications failures.

In any organization, it is essential that three avenues of communication be maintained:

a. downward communication, which allows for a chain-of-command, downward flow of orders and influence;

b. upward communication, which allows subordinates to focus attention on problems developing at their level, thus fostering ego, self-expression, and initiative within the lower echelons;
c. horizontal communication, which allows information to flow at equal levels of authority.

BUDGETING

At the shipboard level, the budgeting process is one of presenting the departmental monetary needs to the commanding officer and justifying them. The basic problem is that a ship never receives all the funds it desires. Therefore, unfortunately, the departments are in competition with one another for available funds. The manager cannot use the approach "I think I need"; he must quantify his data and show his need. For instance, presenting historical data showing quarterly costs of lubricating oil, replacement parts, and other expenses is a good way to justify a quarterly operating target for budget purposes.

Many pages could be written on formal budgeting procedures, but such knowledge is not considered essential for the shipboard duties of the personnel for whom this Guide is intended.

HUMAN FACTORS

Although the authors have, for reasons of clarity, devoted this separate section to this topic, it is actually intertwined among the major management functions.

Motivation and individual human needs have been discussed previously. However, a manager's role in human relations becomes a task of motivation within the framework of a group of individuals.

First, there is the matter of participation. Participation has been defined as the mental and emotional involvement of a person in a group that encourages him to contribute to organization goals and to share responsibility in them. When a person is given the opportunity to use his own initiative and creativeness, participation motivates him to contribute toward the objectives of an organization.

Not only does an individual function as an individual aboard ship, but he also functions as a member of informal social groups that affect the internal workings of the organization. Small groups always exist in a shipboard environment. This is because informal groups naturally develop whenever people work closely together. These men see each other frequently; at work, on the mess deck, and in the machinery spaces. They discover common interests and exchange ideas, and thus form spontaneously into a group, from which one or two individuals emerge as leaders.

Most people get many of their day-to-day human satisfactions from such groups: sociability, inner security and a sense of personal worth, a sympathetic ear for trouble, aid on the job, and some protection through a united stand against unreasonable pressure from superiors. In time, these satisfactions can develop considerable group cohesiveness. Each person normally belongs to several small informal groups, but those based on daily work relationships are likely to be the strongest and most enduring.

It is typical of these social groups to put pressure

on members to conform to group standards and group routines. Restricting production output to a standard set by the group is a common practice, and management can do little to counter it. The group can also influence an individual's feelings toward his superiors. It can influence his feelings about such matters as safety, accuracy in report-keeping, reduction in waste, and work performance. Even when a member has firsthand evidence contrary to group sentiment, he may still accept group judgment rather than "go it alone."

Since the behavior of small groups has an impact on many facets of management, particularly motivation, communication, and leadership, it is essential that managers try to harmonize social groups with the formal organization. The question, then, is to determine how this harmony is to be achieved.

The first step in dealing with informal group actions that do not fit into the formal structure should be to assume that the group standards and behavior are good. It is not necessary to continue on that assumption once investigation has shown that the group's informal action is detrimental to organizational requirements.

A second possibility for harmonizing the practices of small groups with formal organization is to establish an organizational structure that encourages social groups of a type inclined to support, rather than conflict with, the aims of the department. This can be accomplished by using the Task-Team concept in which each team is given the full facts on the task to be accomplished.

Finally, there is the influence of the small-group

clique within the shipboard organization. A clique differs from most informal groups in that its members are spread throughout several different departments. They may never see each other—but—they are united through common support for some "cause." This cause could be a belief concerning policy or even a social philosophy such as racial integration. A clique may promote itself by getting its members appointed to influential positions; by proselyting in various departments and informal groups; by openly advocating its cause whenever it is relevant to the solution of a current problem; and by eagerly carrying out plans that make their cause look good.

Cliques are undesirable if they disrupt departmental programs. When a manager finds a vigorous clique in his group, he should find out what it stands for. Some cliques can be used to sponsor changes in goals or policy and to promote new plans. Others produce a deleterious effect on the organization and should be broken up. The latter type can be established when people

1. fail to accept organization objectives
2. believe standards of performance are set too high
3. believe measurements are inaccurate
4. dislike unpleasant facts
5. find social pressures run counter to organization controls.

From the material contained in this section, it can be concluded that social relationships in an organization are not based on crisp decisions that remain static once they are made, but are based on a series of continuing personal actions and reactions over a long period of time. A manager who understands the social forces at work can design his organization, his planning, controlling, and leadership to use personal social behavior forces for the organization's good.

SUMMARY

Management is defined as the process by which an effective organization directs actions toward common goals. Organizational effectiveness depends on good communications, flexibility, creativity, and genuine psychological commitment, and those characteristics are obtained by

1. recruitment, selection, and training practices that stimulate, rather than demean, people
2. effective group action
3. good leadership in setting goals.

This chapter is intended to give the young shipboard engineering officer an insight into current management practices and to provide broad suggestions for implementation where applicable to the Engineering Department. Complex management applications and procedures, and the numerous quantitative techniques employed in large industrial and high-level governmental agencies have been omitted to prevent further confusing the issue.

CHAPTER 5

The Engineering Department Duty Officer in Port

On balance it may seem like overkill to devote an entire chapter to discussing the Engineering Department duty officer in port (EDDO). But this chapter is being written precisely because the significance of the EDDO is so often underestimated. In small ships, the EDDO may be the senior enlisted man in the Engineering Department duty section. In large ships, such as a cruiser or carrier, he may be a senior lieutenant or possibly a lieutenant commander. Whatever his rank, his basic function and responsibilities are pretty much the same. So, let us identify the EDDO.

The EDDO is the direct representative of the engineer officer on board in the duty section. He is responsible for the proper execution of all Engineering Department functions. Whether he be an officer or an enlisted man this responsibility is the same. It is appropriate here to quote Navy Regulations on the subject of watch officers and the standing of a day's duty.

ESTABLISHMENT OF WATCHES

1. The commanding officer shall establish such watches as are necessary for the safety and proper operation of the command.

2. On board ships the watch of the officer of the deck and of the engineering officer of the watch, shall be regular and continuous, except as hereinafter provided.

3. In ships not underway, the commanding officer may, at his discretion, and subject to such restrictions as may be imposed by a senior in the chain of command, authorize the standing of a day's duty in lieu of the regular and continuous watches described above. [*Navy Regulations,* 1948, Article 1002.]

ASSIGNMENT OF WATCH OFFICERS

1. Subject to such restrictions as may be imposed by a senior in the chain of command, or by these regulations, a commanding officer may assign to duty in charge of a watch, or to stand a day's duty, any commissioned or warrant officer who is subject to his authority and who is, in the opinion of the commanding officer, qualified for such duty.

2. Marine officers below the grade of major may be assigned to duty as officers of the deck in port. Those Marine officers on the junior watch list may stand junior officer watch at sea.

3. At times when the number of commissioned or warrant officers qualified for watch standing is reduced to an extent which may interfere with the proper operation of the command or may cause undue hardship, the commanding officer may assign to duty in charge of a watch, or to stand a day's duty, subject to such restrictions as may be imposed by a senior in the chain of command, or by these regulations, any petty officer or noncommissioned officer who is subject to his authority and who is, in the opinion of the commanding officer, qualified for such duty. [*Navy Regulations,* 1948, Article 1003.]

GENERAL DUTIES OF WATCH OFFICERS

1. An officer in charge of a watch shall be responsible for the proper performance of all duties prescribed for his watch, and all persons on watch under him shall be subject to his orders.

2. He shall remain in charge and at his station until regularly relieved. He shall scrupulously obey all orders and regulations and shall require the same of all persons on watch under him. He shall instruct them as may be necessary in the performance of their duties, and shall insure that they are at their stations, attentive, alert and ready for duty. He shall endeavor to foresee situations which may arise, and shall take such timely and remedial action as may be required.

3. At all times he shall present and conduct himself in a manner befitting his office. His orders shall be issued in the customary phraseology of the service.

4. He shall promptly inform the appropriate persons of matters pertaining to his watch which they should know for the proper performance of their duties.

5. Before relieving, he shall thoroughly acquaint himself with all matters which he should know for the proper performance of his duties while on watch. He may decline to relieve his predecessor should any circumstance or situation exist which, in his opinion, justifies such action by him, until he has reported the facts to and received orders from the commanding officer, or other competent authority.

6. An officer standing a day's duty shall insure, by personal attendance and frequent inspections, that the duties prescribed by these regulations or by the commanding officer are properly performed. [*Navy Regulations,* 1948, Article 1005.]

It is highly advisable that the officers and men who stand duty as Engineering Department duty officer in port be familiar with the above regulations.

One of the most common mistakes made by an EDDO is failing to realize that he has responsibility for functions associated with things outside his specialty area. For example, the chief boiler technician standing duty as EDDO will conscientiously perform his duties with regard to the boiler and main machinery spaces. But will he perform equally conscientiously when the machine shop has a problem, or the electricians are rigging some special lights, or a head overflows? This same tendency occurs in officers as well. The auxiliary division duty officer may do a smooth job in the machine shop, but will he take an equal interest in what boilers are on the line or what generators are being lighted-off or secured?

It is important to ensure that personnel who stand EDDO duty receive sufficient cross training and motivation to function effectively in all departmental areas.

QUALIFICATIONS OF THE EDDO

In ideal circumstances the Engineering Department duty officer in port would be a fully qualified engineering officer of the watch underway. The reasons for this are apparent. Should a casualty occur in the steaming spaces, the EDDO should be able to competently handle the situation. Furthermore, he should be capable of supervising the lighting-off of the engineering plant under both routine and emergency conditions. Since ideal circumstances do not always prevail some compromises frequently have to be made. If there is a senior petty officer in the duty section who is a qualified EOOW or top watch stander, then the duty officer may exercise administrative control while relying on the qualified man to handle a casualty in the spaces. Even with this compromise, the EDDO should have a general knowledge of the plant layout, problems involved in lighting-off the plant, and the evolutions required to effectively run the Engineering Department duty section.

KEEPING INFORMED

It is frustrating for the command duty officer or the officer of the deck to contact the EDDO on a question pertaining to the engineers, only to find that he is totally unaware of the situation. The EDDO should insist that his subordinates keep him informed about evolutions that are taking place. If the personnel in the Repair Division are performing a welding job in a space, the EDDO should know about it. If the boiler tenders have a boiler out of commission for fireside cleaning, the EDDO should be aware of that. Before repair tasks of any nature are performed—and particularly where the safety of personnel or equipment is involved—the EDDO should give his approval.

The EDDO should make periodic rounds of the departmental spaces to keep himself informed and to keep his personnel on their toes. Spaces have been flooded, fires started, and accidents have occurred because of a lack of attentiveness on the part of the EDDO.

CASUALTY CONTROL IN PORT

Some of the most disastrous casualties on record have occurred while ships were steaming auxiliary or on cold iron in port. The reasons are clear. At those times there simply are not as many qualified watch standers in the duty section as there are at other times.

During stand-down periods after a deployment, or during a holiday season when a large percentage of personnel are on leave, the situation becomes even more critical. When schedule and facilities permit, it is wise to go "cold iron" during these periods. Even then, constant vigilance must be maintained to guard against fires and flooding, but the danger of melting-down a boiler or wiping the bearings on a ship's service turbo-generator is certainly reduced.

In addition to the lack of qualified personnel there are other factors that creep into the area of casualty control in port. The men in the oncoming duty section may not be as alert as they should be, or they may not "relieve the watch" properly. Furthermore, they can get lulled into a false sense of security with the steady steaming routine. A ship's electrical and steam load has a tendency to remain fairly constant for long periods of time in port. It is far different from the constantly fluctuating loads which occur in underway fleet operations. As a consequence, there may be an attitude of letting things ride and, when a sudden demand arises, the system fails to respond rapidly and a casualty is in the making. Some common casualties that are more likely to occur in port than at sea are:

1. Loss of feed suction as a result of attempting to take on make-up feedwater from an empty feed tank
2. Loss of fuel suction as a result of failure to shift fuel oil suction on time—or shifting suction to an empty fuel oil service tank
3. "Tripping-out" of generators because of a sudden power surge from large equipment, such as

gun mounts, being lighted-off
4. Oil spills caused from transferring fuel oil without thoroughly checking out the system lineup
5. Flooding of idle spaces
6. Fires in idle spaces
7. Major steam leaks because of improperly wired and tagged valves under repair.

A major complicating factor is the large amount of repair work that is usually taking place when a ship is in port. Pumps are being overhauled, steam leaks repaired, cutting or welding operations taking place, or valves being pulled for repair. Since most of these things do not have to be contended with at sea, the in-port potential for casualties is much higher.

PLACING EQUIPMENT OUT OF COMMISSION

One of the most demanding tasks the EDDO must cope with is placing machinery or equipment out of commission. It is a task not to be taken lightly—for, if it is, loss of life or damage to major machinery can result. NavShips *Technical Manual,* Chapter 9480, gives very good guidance on piping-system repairs. The following provides an example of the kind of information contained in that chapter:

PRECAUTIONS TO BE TAKEN BEFORE BREAKING LINE AND VALVE BONNET JOINTS

1. Before breaking a line or valve bonnet joint, or cutting into a line, ensure the following:

a. Valves isolating the section are shut and secured in such a manner that they cannot be opened accidentally either locally or remotely. The valves should be locked or wired shut and a warning tag attached.

b. The line is completely drained and there is no pressure on the line.

c. Precautions are taken to prevent fire or explosion from flammable liquids.

d. Adequate ventilation is provided.

2. In breaking flanged joints, two diametrically opposite securing nuts should remain tight while the remainder are slackened. These two nuts shall then be slackened sufficiently to permit breaking the joint. After the joint is broken and the line or valve is proved to be clear, all nuts may be removed.

3. Before working on any piping system a survey should be made to assure that under no condition can any liquids splash on exposed electrical equipment. If there is any possibility of splashing a switchboard or other electrical equipment, the following precautions shall be taken:

a. The electrical equipment should be de-energized and covered with a waterproof material.

b. If it is not possible to de-energize the electrical equipment before work on the piping system begins, it should be completely covered by a rubber sheet or other nonconductive waterproof material. Do not restrict ventilation to the point that equipment will overheat.

c. Prior to repairing valves or piping, the piping should be opened away from the electrical equipment, at a lower level if feasible, to ensure that the line is completely drained and unpressurized.

A small ship may be able to function without a formal system for placing equipment or machinery out of commission, but to a large ship it is indispensable. Without a system, disaster is going to strike sooner or later.

What is meant by a "formal" system for placing machinery out of commission? In effect, it is simply a system whereby written permission via the chain of command within the department is followed, starting with the man in charge of the job and ending with the cognizant engineering assistant (main propulsion assistant, damage control assistant, or electrical officer). In some cases, the chief engineer will want to be included as well. After working hours the Engineering Department duty officer has the final authority for placing cognizant pieces of equipment or machinery out of commission. Mere mention of the words *written permission* may conjure up visions of more paper work, but it does not really amount to much extra paper and it is well worth the effort expended. The written-permission system serves two functions: it keeps those who need to know informed of what is going on in the department; and it provides a margin of safety that cannot be achieved by the more informal word-of-mouth approach.

The format for the written permission to place equipment out commission can be designed to fit the needs of any ship, large or small. As a minimum, it should

1. specify the equipment to be worked on
2. identify other units or systems that might be affected and the isolating valves that will be secured
3. estimate time of completion
4. include a note on the safety precautions that should be observed
5. estimate date for placing the equipment back in commission
6. include an approval section for signature by persons in the chain of command.

The chain of command that was advised of equipment being placed out of commission should be informed about its being placed back in commission. In conjunction with the written permission, a small pocket book of system diagrams is very helpful. Using the two together provides a means of double-checking the isolating valves to be sure that the equipment is safely isolated and that other systems will not be adversely affected by the closing of the isolating valves. There have been occasions when engineers attempting to work on an auxiliary steam line have unwittingly cut off steam to the galley at a critical time in meal preparation, thus inconveniencing the crew and infuriating the supply officer. At other times a critical system has been isolated without notification, with the result that the electrical load was lost, or worse. So, it is important from both a safety and an operational standpoint that each man should know exactly what he is doing when he places a piece of machinery out of commission. The subject of placing equipment out of commission is discussed in more detail in Chapter 11.

Placing machinery back in commission is perhaps even more critical. Here is a situation where a system that has been isolated has to be repressurized, and any opening in an idle line can have disastrous consequences. Sometimes, an isolating valve will be closed and tagged two or three times if it happens to control pressure to more than one branch of piping and the other branches are being worked on. Such a tag should be clearly marked

DO NOT OPEN without permission
of the undersigned
(Signature)

Obviously, a valve cannot be safely opened until *all* tags have been removed and *all* repair work has been complcted.

EDDO DUTIES WHEN GETTING UNDERWAY

The EDDO has some special functions to perform should he have the duty the day before his ship gets underway. He should be planning ahead in order to reduce to a minimum the incidence of last minute confusion. Among other things he might ask himself are:

1. Has the gyrocompass been lighted-off far enough in advance to be settled out prior to underway time?
2. Are there sufficient movies on board for the next at-sea period?
3. Has the lighting-off schedule been promulgated and the watch set up?
4. Are there any major pieces of machinery or equipment out of commission which might prevent the ship from getting underway?
5. If special decorative lights are rigged, has it been arranged for them to be taken down prior to getting underway?
6. Is the ship topped off on potable water?
7. How much feedwater and fuel oil is on board?
8. Are there any last-minute details that need cleaning up before the ship gets underway?
9. Is there anything the chief engineer should know about before the ship gets underway?
10. If outside telephones are rigged, has someone arranged to disconnect them prior to getting underway?

11. Have the running lights been tested?
12. Have drafts fore and aft been taken?

In a well-run ship most of these things happen automatically, and one of the reasons they happen automatically is because the officers and men who stand EDDO take their jobs seriously. One of the marks of a well-organized Engineering Department is that it runs well when the "boss" is absent. The EDDO can and should make a major contribution to that end. A typical Engineroom Warming-up Schedule and a Lighting-Off Schedule are shown in Appendix B, along with notes prepared by ComCruDesPac Engineering School, San Diego. These should be helpful as a guide for starting and operating the plant.

CHAPTER 6

The Engineering Officer of the Watch Underway

STATUS, AUTHORITY, AND RESPONSIBILITY

> The engineering officer of the watch is the officer on watch in charge of the main propulsion plant of the ship, and of the associated auxiliaries. He shall be responsible for the safe and proper operation of such units, and for the performance of the duties prescribed in these regulations and by other competent authority. [*Navy Regulations,* 1948, Article 1025.]

The prominence of the engineering officer of the watch (EOOW) varies within ships of the same type and between ships of different types. On large ships, where the engineering plant is complex, experienced officers in the Engineering Department are almost always excused from deck watches in order to assume the duties of EOOW. On small ships, because of limited officer personnel allowance, officers in the Engineering Department must, of necessity, assume the duties of officer of the deck, junior officer of the deck, and CIC watch officer underway. Thus, an experienced and dependable senior petty officer in the machinist's mate or boiler technician rating is usually assigned duty as the EOOW.

Rank is by no means the most important consideration in qualifying as an EOOW. Primary requirements are a thorough knowledge of the plant and a demonstrated competence in watch-standing. The EOOW must possess the knowledge and experience that will enable him to act quickly and confidently in any situation.

Some engineer officers, in addition to appraising an individual's demonstrated watch-standing ability, may require prospective EOOWs to take a written examination. This procedure has some merit in that

1. it lets a new man coming on board know that he must measure up to a specific standard.
2. it establishes a common set of requirements which remain effective in spite of frequent personnel transfers.
3. it spotlights areas that need further attention.

Even the most highly qualified EOOW will be severely handicapped if he does not have a properly equipped watch station to work from.

WHERE THE WATCH IS STOOD

One of the riddles of World War II ship design was the lack of emphasis placed on providing naval vessels with a satisfactory main propulsion control station. Many ships built in the 1940s provided few, if any, facilities for the engineering officer of the watch. However, this problem has been treated in a much more realistic manner in new construction and conversion of ships in recent years. As a result, ships in today's fleet offer an assortment of engineering control stations ranging from the homemade telephone-booth variety to the expansive, air-conditioned, sound-isolated control room. Even the names differ: main control, main engine control, and central control are samples of the most widely accepted titles.

Vast differences in equipment exist at control stations even on combatant ships. For example, the EOOW on a standard 2,200-ton destroyer stands his watch at the throttle board of the forward engineroom. This "main control" position usually satisfies the minimum physical requirements of having

1. a gage board for the forward plant and selected vital indicators for the after plant
2. status boards
3. essential communications circuits
4. throttle controls.

In such a reasonably compact engineering plant, outfitted with minimum control arrangements, it is easy to understand why the role of the EOOW is assumed by a highly qualified engineroom enlisted top watch.

Taking the other extreme, a guided-missile cruiser (CG) has a quiet, air-conditioned, central control station located in a compartment completely removed from the machinery spaces. Remote-reading display boards present a convenient array of essential data for propulsion and electrical-distribution control. Various alarm and communication circuits provide a responsive network for the rapid collection and dissemination of information. This control center furnishes the EOOW with the means to make sound, reasoned decisions, no matter how chaotic the situation.

Regardless of where the watch is stood, the EOOW must ultimately rely on the members of the watch team to perform their duties in a professional manner. The command to "cross connect the plant" means nothing unless at the other end of the communications link there are alert men who know which valves to turn. Of course, the larger the ship the more complete this reliance must be.

Imagine, for example, the task faced by the engineering officer of the watch on a *Midway*-class attack carrier during a busy watch. He must maintain positive control over twelve boilers, four main engines, and eight ship's service turbo-generators—all remotely located from him and from each other; even the main feedpumps and the deaerating feed tank are located in a pumproom separate from the engineroom. In addition he must coordinate routine housekeeping chores, such as pumping bilges and blowing tubes, supervise placing various units of machinery in and out of commission, maintain the Engineering Log and the Engineer's Bell Book, be ready to supply steam to the catapults, and respond instantly to major speed changes when flight operations are underway. If the EOOW is to function effectively, he certainly must understand, and work in harmony with, the existing watch organization.

ORGANIZATION OF THE WATCH

The details of watch organizations vary widely from ship to ship and are mainly dependent on the arrangement of the various components of the main power plant. The most important consideration is: "Are there enough people, properly employed, to do the job effectively and safely?" The EOOW should be alert to spot weaknesses in the watch organization and bring them to the attention of competent authority.

RECORDS AND LOGS

> The engineering officer of the watch shall insure that the engineering log, engineer's bell book, and prescribed operating records are properly kept and that the bell book and other operating records are signed by individuals who have knowledge of the orders given and executed. On being relieved, the engineering officer of the watch shall himself sign the engineering log for his watch. [*Navy Regulations,* 1948, Article 1031.]

LEGAL RECORDS

The required legal records and what they imply, substantiate the need for proper record-keeping as it relates to the engineering watch. For convenience, the applicable article from NavShips *Technical Manual* Chapter 9004, is quoted here in its entirety:

> 1. The Engineer's Bell Book (NAVSHIPS 116) is a record of events made at the time they occur. The instructions for the use of the Engineer's Bell Book provide that the propeller speed recorded must be the r.p.m. required by an order directing a change in propeller speed, and not the r.p.m. which results from the order.
>
> a. Before going off duty, the engineering officer of the watch must sign the Bell Book in the line following the last entry for his watch and the next officer of the watch shall continue the record immediately thereafter. In machinery spaces where an engineering officer of the watch is not stationed, this record must be signed by the senior petty officer of the watch. Alterations or erasures are not permitted. An entry which is incorrect must be corrected by drawing a single line through it and making the correct entry on the following line. Such deleted entries must be

initialled by the engineering officer of the watch, senior petty officer of the watch, or O.O.D. (in the case of ships and craft equipped with controllable reversible pitch propellers), as appropriate.

b. The records for each throttle control station for each day must begin with a new sheet, and the day's records for all stations must be clipped together and filed as a unit. The Engineer's Bell Book is not required to be maintained when the propeller shafts are not turned over either by jacking or with the propulsion machinery.

c. The engineer's Bell Book sheets must be preserved as a permanent record on board except in obedience to a demand from a Naval Court or Board, or from the Navy Department. In that case a copy, preferably photostatic, of such parts of the record as may be sent away from the ship, shall be prepared and certified as a correct copy by the engineering officer for the ship's files. Bell Books may be destroyed 3 years after date of last entry. When the ship is stricken, the current books must be forwarded to the nearest Naval Records Management Center.

2. The Engineering Log (NAVSHIPS 117) is applicable for both surface ships and submarines and is a daily log showing the data described in the instruction printed on this form.

a. The original record, neatly prepared in ink or pencil, is the legal record. The remarks should be prepared and signed by the officer-of-the-watch on the day before leaving his station or being relieved. Any errors should be overlined and initialed by the person preparing the original entries.

b. The engineer officer must verify the accuracy and completeness of the entries and sign the log daily. The commanding officer must sign the log on the last calendar day of each month, and on the date of relinquishing command.

c. The Engineering Log must be preserved as a permanent record on board except in obedience to a demand from a Naval Court or Board, or from the Navy Department. In that case a copy, preferably photostatic, of such sheets as may be sent away from the ship shall be prepared and certified as a true copy by the engineering officer for the ship's files. Engineering Logs may be destroyed 3 years after date of last log entry.

When a ship is stricken, the current logs must be forwarded to the nearest Naval Records Management Center.

OPERATING RECORDS

The prescribed operating records are discussed in Chapter 3.

SIGNIFICANCE OF RECORDS

The sequence of events leading up to a collision or grounding can be traced and verified from a well-kept Engineering Log and Engineer's Bell Book. Action taken to combat and control fires, flooding, and machinery derangements can, if recorded accurately, be of significant value in improving damage and casualty control techniques.

From a personal point of view, the legal engineering records provide a reasonable measure of protection for the individual involved in an incident, whether he be the man on the throttle, the EOOW, the officer of the deck, or the commanding officer.

For example, imagine you are the EOOW standing a quiet Sunday afternoon watch underway in the Mediterranean. The engineroom top watch informs you that he hears an unusual noise in the low-pressure turbine of number two main engine. What do you do? Let us assume you notify the officer of the deck and the engineer officer immediately, and proceed to take all of the correct actions as specified in your ship's casualty control book and NavShips *Technical Manual.* In short, you do an absolutely superb job of controlling the casualty—except, you fail to enter the chronological sequence of events in the Engineering Log. On return to port, the turbine casing is lifted only to find the low-pressure turbine blading has suffered major damage. A board of investigation is convened. The most valuable piece of evidence you could have produced would have been a complete and accurate account of the incident recorded in the Engineering Log.

INSPECTION AND OPERATION OF MACHINERY

> The engineering officer of the watch shall cause frequent inspections to be made of the engines, boilers, and their auxiliaries; and shall insure that prescribed tests, methods of operation, and instructions pertaining to the safety of personnel and material are strictly observed. [*Navy Regulations,* 1948, Article 1030.]

RESPONSIBILITY AND HONESTY

Every seasoned engineer can recall potentially disastrous situations caused by personnel reporting incorrect information as to work accomplished. The compartment that *almost* flooded, or the pump that *nearly* ran out of lube oil are all too common occurrences on board ship. Why does this kind of near-accident happen? There are as many different reasons as there are people.

A young fireman may be terrified at the thought of someone finding out that he forgot to inspect the leak-off on the stern tube of the port shaft during his last routine inspection of the shaft alleys. In fact he may be more worried about someone finding out, than he is about the shaft alley flooding! He is concerned about his own well-being and has not yet developed a sense of responsibility for his ship and shipmates. If detected in time, this man's problem can be corrected through effective leadership.

The recurring tendency to "gundeck" readings must be recognized and constantly guarded against. Many engineer officers are painfully familiar with the beautifully inscribed bearing temperature log which presents a neat row of figures marching straight down the page, steady and unchanging until the bearing suddenly and mysteriously "self-destructs"!

Generally, readings are gundecked for one of two reasons, (1) laziness, and (2) ignorance. The first reason is inexcusable and can be dealt with at captain's mast. The second reason can be eliminated through proper education. The whole reason behind taking temperature and pressure readings on a fixed schedule is to rapidly pinpoint abnormal conditions. As an absolute minimum training program, a man assigned to take readings should be shown samples of unusual conditions and given a complete explanation of the possible causes and effects of each. It is vital that each watch-stander be educated as to *why* meticulous performance of his duty is necessary. A sense of responsibility will follow naturally.

Until the day comes when ship propulsion systems are fully automated, the innumerable details of operating the plant manually will continue to fall to the "man on watch." In the engineering spaces particularly, relatively junior men are often entrusted with the sole responsibility for inspecting, operating, and maintaining the major components of the plant. It is imperative that every man who stands the watch "down below" be indoctrinated in the importance of absolute and unwavering honesty. There are many stories concerning major machinery damage, fires, floodings, injuries, and loss of life in the engineering spaces of naval ships. The greatest tragedy of all is that many such incidents could have been foreseen and prevented had the watch-standers involved held to a creed of complete honesty and integrity.

Nothing sums up this section better than the words of Admiral David L. McDonald, U. S. Navy (Retired), Chief of Naval Operations from 1963 to 1967:

> Always tell the truth and you never have to remember what you said. . . . [*U. S. Naval Institute Proceedings,* January 1970.]

REPORTS AND COMMUNICATIONS

REPORTS TO THE ENGINEERING OFFICER OF THE WATCH

> The engineering officer of the watch shall be promptly informed of any engineering work or change of disposition of machinery which may affect the proper operation of the plant or endanger personnel, or which is required for entry in the record of his watch. [*Navy Regulations,* 1948, Article 1029.]

REPORTS BY THE ENGINEERING OFFICER OF THE WATCH

> The engineering officer of the watch shall report promptly to the officer of the deck and the engineer officer any actual or probable derangement of machinery, boilers, or auxiliaries which may affect the proper operation of the ship. [*Navy Regulations,* 1948, Article 1028.]

COMMUNICATIONS WITH THE BRIDGE

> The officer of the deck is the officer on watch in charge of the ship. He shall be responsible for the safety of the ship and for the performance of the duties prescribed in these regulations and by the commanding officer. Every person on board who is subject to the orders of the commanding officer, except the executive officer, and those other officers specified in Article 1009, [Command Duty Officers and Navigator] shall be subordinate to the officer of the deck. [*Navy Regulations,* 1948, Article 1008.]

In his association with the officer of the deck, the engineer sometimes assumes that the bridge neither knows nor cares about what is taking place in the engineering spaces. In a well-run ship such an assumption would have to be counted wrong by its very nature. The problem of communications with the bridge can be examined from two totally different points of view.

It is a commonly accepted principle that every officer of the deck needs at least an elementary understanding of shipboard engineering fundamentals if he is to stand his watch properly. As a minimum, he should be thoroughly familiar with the meaning of those engineering terms which are most routinely exchanged between the OOD and the EOOW. The officer of the deck should be an expert on engine order telegraph (EOT) procedure, including emergency bell signals and action to be taken in case the EOT malfunctions. Finally, in an engineering sense, he should know the cause-and-effect reactions which are set in motion by sudden major changes in the ship's speed or direction.

On some ships prospective OODs are required to

stand a series of familiarization watches in the engineering spaces prior to qualification as deck watchstanders. These watches can be particularly informative if they are stood during a period when the ship is maneuvering with the engines, such as when leaving or entering port. Fifteen minutes spent in a fireroom watching some fast fuel-burner changes and forced-draft blower adjustments can be more educational to the uninitiated than hours of lecturing.

It would seem logical that if the officer of the deck is expected to know something about engineering, the EOOW should know something about being an OOD. This is a more difficult goal to achieve, and is frequently the root cause of any lack of coordination which might exist between main engine control and the bridge. The EOOW is usually a more restricted specialist than his deck watch-standing counterpart. As a result the EOOW does not always have the "big picture," and he thinks of what is taking place around him only in terms of how it affects his particular specialty. For example, a machinist's mate, first class, standing EOOW on a destroyer may know very little about the OOD's problem of keeping the ship on station in a formation. In a similar sense, an engineering-oriented limited duty officer could be somewhat restricted in his knowledge of bridge functions. Because of these special circumstances, the engineer officer should take the initiative in establishing and maintaining high standards of competence and effectiveness in his department's communications with the bridge.

A few simple rules, and a knowledge of the reasons for their existence, will contribute to smoother coordination between engineering control and the bridge.

1. *Think before you speak.* Sometimes a discreet inquiry via the sound-powered phone circuit can save much embarrassment. It can prevent such errors as
   a. requesting permission to blow tubes on the steaming boilers of an aircraft carrier during the launch or recovery of aircraft.
   b. requesting permission to pump bilges or dump trash as the ship is approaching the pier to tie up.
   c. requesting permission to secure the main engines before the ship is safely alongside the pier.

   When such errors become the rule rather than the exception, the competence of the entire Engineering Department becomes suspect in the eyes of the officers of the deck and the commanding officer. It is a wise EOOW who, when he talks to the bridge, assumes he is speaking directly to the commanding officer—more often than not he is!

2. *Be precise with your words.* Don't assume that the OOD knows about everything that is going on in the engineering spaces. He may be deeply involved in maneuvering the ship or handling some administrative matter. A sudden announcement from below that "the plant is cross connected," or that "number two boiler has been secured," might leave him thoroughly confused. An informative statement such as "completed cross connecting the plant at time 2034, in accordance with the engineer officer's night

orders," tells the OOD exactly what has been done and why.

3. *Anticipate the OOD's need for information.* It is a distinctly frustrating experience to be standing a bridge watch and hear a vaguely muffled report to the effect that number one boiler has just suffered a low-water casualty, followed by several minutes of silence as the ship lurches slowly to an uncertain stop. It is not always easy for the EOOW to control a casualty and keep the bridge informed at the same time, but it is a skill he must develop, lest he someday find himself in the awkward position of saving the engine while losing the ship.

The requirement to keep the bridge informed with regard to casualty and damage control is covered extensively in the chapters on those subjects. In the meantime, it is worthwhile to remember that a ship spends the great majority of her lifetime under routine and peaceful conditions. Effective communications with the bridge can go a long way toward keeping those conditions as they should be—routine and peaceful.

DIRECTING AND RELIEVING THE EOOW

> The engineer officer, or, in his absence, the main propulsion assistant may direct the engineering officer of the watch concerning the duties of the watch, or may assume charge of the watch and shall do so should it in his judgment be necessary. [*Navy Regulations,* 1948, Article 1026.]

There is a clear implication in the preceding quotation that the engineer officer and the main propulsion assistant are the most highly qualified EOOWs on the ship. In reality, of course, this cannot always be—but it is certainly a worthy and necessary goal to strive for. Lack of qualifications in no way reduces the ultimate responsibility.

There is the story of an officer who received orders to be the engineer officer of a particular ship. Obtaining copies of the piping diagrams for that ship's plant, the prospective engineer studied them extensively before reporting for duty. As a result he was roughly familiar with the plant the day he stepped on board. Building on that background he rapidly became the most knowledgeable and respected officer on the ship in matters of engineering. His officers and men appreciated his steady hand and easy competence under every conceivable circumstance. To keep himself up to date on the routine in the engineering spaces, he occasionally stood in as EOOW. He was thoroughly familiar with the most intricate details of casualty control and lighting-off or securing the plant. In addition to all this, he administered the Engineering Department of an attack carrier—a total of more than 900 officers and men. He was a seagoing engineer of the highest calibre and an inspiration to all who knew him.

In the end, every officer who serves a tour or two of duty in engineering will inevitably be required to face his moment of truth—that moment when the lights go out, the "blowers" fall silent, and the ship goes dead in the water. At that moment, *he* must find the solutions and *he* must make the decisions that could make the difference between losing or saving a man's life, or perhaps even, the life of the ship.

CHAPTER 7

Casualty Control

This chapter discusses the philosophy and practice of basic casualty control under various headings. Since the design of a ship's engineering plant is beyond the province of most seagoing engineers it will not be discussed. However, this omission does not imply that sound design is unimportant. On the contrary, proper marine power plant design and construction is fundamental to casualty-free operations. The ultimate goal of casualty control from the engineering standpoint is to retain a maximum of propulsive and auxiliary power when battle damage is sustained.

MISSION OF CASUALTY CONTROL

Engineering casualty control is concerned with the prevention, minimization, and correction of operational and battle casualties to the machinery, electrical, and piping installations. The mission of casualty control is the maintenance of all engineering services in a state of maximum reliability under all conditions of operation. The first objective under this mission is the effective maintenance of propulsion, auxiliary and electric power, lighting, interior and exterior communications, fire control, electronic services, ship control, supporting vital systems for main, auxil-

> iary, and electric power, firemain supply, miscellaneous services, such as heating, air-conditioning and compressed air. Failure to provide all normal services will affect the ship's ability to perform effectively as a fighting unit, either directly by reducing its mobility, offensive and defensive power (including the ability to control fires, flooding and hull and armament damage), or indirectly by reducing habitability and thereby reducing personnel morale and efficiency. The second objective is the minimization of personnel casualties and secondary damage to vital machinery, since these factors contribute in a large degree to the successful and continued accomplishment of the first objective. [NavShips *Technical Manual,* Chapter 9880.]

It is apparent from the above quotation that engineering casualty control is a tremendously complex operation of overwhelming importance to the ship and to all who sail in her. Fortunately, the simplifying factor of similarity between systems reduces the problem to one of manageable proportions. For example, a steam propulsion plant, regardless of its design operating temperatures and pressures, is still simply a steam propulsion plant. Such plants may vary widely in appearance, method of control, piping arrangement, size, power capacity, flexibility, speed of response, and numerous other characteristics, but the fundamental principles of operation are the same. This same line of reasoning applies to diesel engines, gas turbines, electrical systems, compressed-air systems, air-conditioning systems, firemain systems, and so forth. Once the fundamental operating principles of a system are understood, the specific details applicable to an individual ship can be readily acquired. In theory, naval equipment is standardized in order to reduce the amount of re-learning required when personnel are transferred from one ship type to another; in practice, each ship has variations from the standard.

PREVENTIVE MAINTENANCE

> The basic factors influencing the effectiveness of engineering casualty control are much broader than the immediate actions or routines applied at the time of the casualty. Engineering casualty control reaches its maximum efficiency by a combination of sound design, careful and continued inspections, planned maintenance, and effective personnel organization and training. *Casualty prevention is the most effective form of casualty control.* [NavShips *Technical Manual,* Chapter 9880, Factors Influencing Casualty Control.]

A quality machine that is properly operated and carefully maintained will rarely fail in service. Such careful maintenance can be realized through the effective use of the Standard Navy Maintenance and Material Management (3-M) System, which is discussed in Chapter 11. This system provides a modern, integrated approach to the scheduling, recording, reporting, and managing of ship maintenance actions throughout the fleet.

Experience in the fleet confirms the fact that the incidence of malfunctions such as burned-out pumps, wiped bearings, and shorted electrical equipment can be substantially reduced by conscientious application of the measures outlined in the Navy's Planned Maintenance System (PMS). The PMS is the preventive maintenance component of the total 3-M system.

It is deceptively easy for a ship to reach a condition in which material casualties begin to occur at a greater rate than they can be repaired. This is particularly true in old ships or in ships that are overdue for overhaul, and it means that the engineers find themselves in the position of firemen who must spend most of their time rushing about putting out small fires before they become major conflagrations. An evaporator brine pump burns out, or a fire and flushing pump freezes up, or the meat grinder in the galley strips a gear, or a recently repaired boiler fails to pass a hydrostatic pressure test. When this atmosphere of crisis extends over a long period of time emergency stopgap measures often come to be accepted as standard operating procedures, and the end result is the weakening or loss of any effective preventive maintenance routine that might have existed. It is simply illogical and impractical to expect a crew to expend effort on equipment that might fail when they are faced with a mountain of repair work on machinery that has already failed. This situation is unlikely to develop when a planned maintenance system is effectively organized and carried out on a continuing basis. The struggle to overcome a heavy backlog of casualties and bring a ship back to a reasonable degree of operational reliability is always a long and difficult task. It leaves little time for the luxury of practice casualty control. Even when a desirable level of material readiness is achieved, any significant relaxation of maintenance standards can trigger the beginning of the whole disruptive cycle all over again.

For the remainder of this chapter it will be assumed that an effective preventive maintenance program is being carried out. This, of course, will reduce the number of actual casualties a particular ship may encounter—but it will not eliminate them. Operational casualties due to material failure and human error can and will occur.

KNOWLEDGE AND PRACTICE

The two mandatory prerequisites for effective casualty control in their order of importance are *knowledge* and *practice.* Effective casualty control requires that each man possess a complete and thorough knowledge of all aspects of that portion of the engineering plant for which he is responsible. Furthermore, he should be drilled to a degree that will allow him to move quickly and confidently in any emergency. For instance, engineroom personnel should know the location and function of every major valve in their space. In addition, the more experienced hands should have a reasonably broad knowledge of piping and machinery extending beyond their own position or space. A man attempting to control a casualty in the engineroom should know the effect his actions may have on other portions of the plant. For example, when the engineroom throttleman is applying astern steam to stop a shaft, he must be aware of the effect of his actions on the fireroom. If the throttleman disregards the warning of a pressure gage indicating a drop in main steam pressure, he may drag a boiler right off the line. This will, of course, compound an already serious casualty.

There are innumerable examples of how things can go wrong in casualty control, either practice or real. This is why it is imperative that comprehensive walk-through drills be conducted before an actual active exercise is attempted. It is dangerous and foolhardy to do otherwise.

A STATE OF MIND

Frequently, there is a strong natural reluctance to conduct active casualty-control drills on board ship. The potential for equipment damage seems very real when compared with the somewhat intangible benefits derived from this particular type of training. There is another more subtle element that contributes to this reluctance. For lack of a better definition it can be called, *fear of the unknown*. What will happen if the ship loses all electrical power? What does one do if the master gyrocompass tumbles? What if the emergency diesel generator fails to pick up the load, and steering control is lost? Reluctance concerning casualty-control drills is a symptom that the Fleet Training Group finds on most ships. But week after week they watch ships come and go, and they see an entire crew's attitude gradually transformed from one of initial uncertainty to one of burgeoning self-confidence. This is accomplished through proper casualty-control training procedures. The shipboard engineer's job continues where the Fleet Training Group leaves off. He must continue to ask the embarrassing questions that determine the difference between superficial knowledge and real understanding: for example, how much do the average engineering or bridge watch-standing officers really know about the after steering system on their ships? Too often, the procedure for *loss of steering* consists simply of ringing the alarm and shifting cables. What if that doesn't work? What is the next step? If the after steering compartment is flooded, will the steering machinery still function? How? Is there a manual method for providing hydraulic pressure to the steering gear? How much time is required to accomplish each corrective action?

Wide application of this kind of in-depth probing for knowledge can clear away many of the myths and half-truths that have a tendency to build up around things that are not fully understood.

No doubt the practice of casualty control is a calculated risk, but thorough planning and preparation can reduce that risk to a minimum. The skills and experience gained can also pay big dividends to a crew in terms of pride and confidence in knowing they can handle any situation that may arise. The alternative of having a completely untrained crew that is either too uninformed or too timid to take immediate and positive action when the chips are really down could be dangerous, at best.

Finally, it should be realized that casualty control is not an evolution to be engaged in by engineers alone. It affects and involves in some way almost everyone in the ship. When normal electrical power is lost, the men in the Combat Information Center and on weapons stations must know how to restore power to their equipment as rapidly as possible. Sound-powered phone circuits can become clogged with complaints about equipment that isn't working or a loss of space ventilation—all because operating personnel are not familiar with the power-switching arrangement on their stations. In these respects, every man must be a casualty control man.

ORGANIZATION AND TRAINING OF PERSONNEL

There are three situations that present distinctly

different problems in terms of casualty control organization and training:

1. general quarters or special evolutions, which include condition watches and underway refueling and special sea details
2. normal steaming watches
3. in-port auxiliary watches.

When general quarters has been sounded or special evolutions are underway, every member of the crew is at the station he is best qualified to handle. Therefore the primary training task is to ensure that the crew functions as a smoothly coordinated team.

For the normal steaming situation the problem becomes twofold. First, the underway watch sections must be assigned with a view to spreading the experience level in the most equitable and efficient manner. Second, a training program for each section must be established. The crew's experience level determines just how extensive this training needs to be. On a small ship where a crew has been thoroughly trained at general quarters, the need for watch-section training may be minimal, but on a large ship, with a new crew, a great deal of training may be necessary, and should be conducted as a beginner's course. There is sometimes a tendency, even among the experienced, to subconsciously assign their own individual abilities to their men. A chief engineer who is thoroughly knowledgeable and highly proficient in casualty control might assume that his Engineering Officers Of the Watch (EOOWs) are equally qualified. He may reason that things that are familiar to him are just as familiar to his officers. This is simply not so.

Finally, the most frequently overlooked casualty control situation is the in-port auxiliary watch. In spite of the fact that the main engines are secured, there is still an enormous potential for trouble in an inexperienced, poorly supervised in-port watch. The rush to hit the beach at liberty call should never be allowed to overshadow the best interests or safety of the ship. Elimination of last-minute scrambles for duty exchanges and insistence on equal qualifications for standbys is a good first step. The in-port watch is always the weakest of the steaming watch situations. During leave and holiday periods it can become critical. For these reasons it deserves a good measure of attention and planning. A more detailed discussion of in-port engineering watches is contained in Chapter 5.

COMMUNICATIONS

A consistent set of instructions should be laid down with regard to casualty-control terminology. The precise meaning of each phrase must be common knowledge to every officer and man involved in the operation and control of the plant. For example, although the order to *cross connect the plant* is a common engineering casualty-control phrase, it may mean different things on different ships. On one ship it may mean that all systems such as main, auxiliary, and ship-service generator steam, main feed and condensate, auxiliary exhaust, and so forth, are to be cross connected. On another ship it may apply only to certain pre-designated systems. In this regard, the key valves in each system should be positively checked each time a system is split or cross-connected. Memory should *not* be relied upon in these situations. When time permits, each machinery space should report to the EOOW the identification number of all key valves and whether they are opened or closed. This information should be checked against the system diagram standard or valve display board which is normally maintained at main engine control.

Of course, every casualty-control order cannot be reduced to a simple phrase or two, and it is not always possible to control a casualty *by the numbers.* Each casualty has to be considered in the light of existing circumstances and, then, clear and specific orders must be given to correct or control it. Only experience will point out those areas which lend themselves to the effective application of abbreviated orders and phrases.

It is admittedly difficult, particularly for the men in the affected spaces, to give their undivided attention to controlling a casualty and at the same time report it promptly and correctly. Nevertheless, this requirement must be emphasized over and over again until it becomes automatic. Communications must flow smoothly between all engineering spaces and the EOOW, and between the EOOW and the officer of the deck on the bridge. Sometimes this process will be the ultimate in simplicity, while at other times it may require a rather complicated discussion. The following is an example of effective casualty-control communications in which a minimum of conversation is called for:

> A ship is steaming at standard speed with the crew at general quarters. The captain and the engineer officer have worked out in advance a mutually agreeable

> system for keeping the bridge informed of the approximate maximum speed available at any given time. With standard speed rung up on the Engine Order Telegraph (EOT), a casualty occurs to the main propulsion plant which reduces the maximum speed available to ⅓. The engineers ring up ⅓ on the EOT. The bridge rings up ⅓ and then immediately rings up standard speed again. In this exchange, the engineers have notified the bridge that only ⅓ speed is available while the bridge has acknowledged this fact and reaffirmed their desire for standard speed. When more power is available, the engineers ring up ⅔ on the EOT. Again the bridge matches the ⅔ bell and then immediately rings up standard speed. Finally when the casualty has been corrected, the engineers ring up standard speed (thus matching the order from the bridge) and report to the bridge that they are ready to answer all bells. This method of exchanging signals meets the objective of effective communications with a minimum of conversation.

It is the engineer officer's responsibility to determine the commanding officer's wishes with regard to communications between engineering control and the bridge. Then he must take steps to ensure that those wishes are met effectively.

One of the most vital aspects of good communications in casualty control is *timeliness.* Every attempt should be made to keep all stations informed about impending casualties as far in advance as possible. Even if a casualty does not materialize, little harm can come from being forewarned and forearmed. There have been many occasions when the prompt reporting of a seemingly insignificant malfunction has resulted in the arrestment of a major casualty The report that a space is *having trouble* maintaining a vacuum in the main condenser or holding down the lube oil temperature on a bearing may be the first danger signal of a potentially serious material failure. The EOOW should be alert to detect persistent gage fluctuations and quick to question the top watch of a machinery space that consistently fails to report such incidents. If watch-standers report that their gages are *in error,* the gages should be calibrated, repaired, or replaced.

Before concluding the discussion of casualty-control communications there is one critical requirement that must be recognized. No system of communications, and especially no sound-powered system, can function efficiently unless it is operated by competent personnel. In the sound-powered telephone system, effectiveness is a direct function of the competence of the phone talker. Phone talkers should know what they are talking about and be able to convey a message clearly. Do not fall into the trap of always assigning the least qualified watch-stander to the duty of phone talker. Slack periods of operation can be used to give new men experience on the phones, but it should not become standard policy to assign recruits to phone-talker duty. Any engineer who has ever lighted-off a plant for getting underway or has handled a serious casualty will testify to the fact that excellent phone talkers help tremendously in such situations. On the other hand, care should be taken to recognize and tactfully correct the phone talker who becomes so proficient that he begins to usurp the rights and prerogatives of the watch officer. An outstanding phone talker is a precious commodity on board any ship and he should be treated with the respect he has earned. But he should never be given the impression that he has assumed the duties of the officer or petty officer of the watch.

DUTIES OF THE TOP WATCH IN CASUALTY CONTROL

Whether in port or underway the key man in the casualty control organization is the top watch in each space. In every case it is the top watch who sets the standards of performance for his watch team. A good top watch-stander is like a good football quarterback: he calls the signals and the players run the play with precision because each man knows what he is supposed to do and when he is supposed to do it. Every man who stands an engineering top watch must be instilled with a sense of responsibility for his team. If possible, the men should stand watch in the spaces they are assigned to maintain. They are not only more familiar with these spaces but, perhaps more importantly, they are standing watch with men they know and work with daily. An additional benefit of such an arrangement is that a spirit of competition can be more easily established. A competitive spirit, if properly encouraged, can contribute greatly to morale. The men in the forward fireroom will take justifiable pride in the fact that they can consistently secure a boiler or shift fuel oil suction faster than their peers in the after fireroom. The after engineroom knows that, if given the choice, the chief engineer will invariably select their space for conducting a competitive exercise because they always look sharp and act sharp. The desire to excel should be recognized and rewarded at every opportunity.

BE PREPARED

How does a man acquire the clairvoyance that allows him to anticipate and prepare for what might happen? It is, quite simply, a matter of attitude and experience. One must become a professional pessimist and constantly ask oneself: What would happen if . . . and what corrective action would have to be taken? What would happen if there were a loss of main lube-oil pressure while the ship was engaged in a highline transfer? What if a boiler lost fuel-oil suction or water got into the fuel oil while the ship was maneuvering to come alongside a pier? If, in the middle of a missile shoot, an air compressor tripped off the line—would such a casualty affect the shoot? If a fireroom or engineroom lost all ventilation, what would be the effect on that space and on the ship? If the emergency diesel generator is to be tested, is it done by dropping the electrical load so that all of the automatic devices are cycled? If, while lighting-off the plant for getting underway, the engineroom watch sent cold main condensate to the deaerating feed tank that was steaming auxiliary—what could happen?

The man who constantly asks himself such questions will not only be better prepared to cope with a casualty when it arises but, in many cases, he will be able to prevent their occurrence. Of course, there is no way to ensure that every possible malfunction has been considered in advance, but certainly many potential disasters can be eliminated and the element of surprise can be greatly reduced. In keeping with this philosophy of preparation, the casualty-control manual should be reviewed thoroughly and often. If it is not up to date or if it contains erroneous information it should be corrected, otherwise it will be of little value to the very people who should be reading it.

THE DOMINO EFFECT

The *domino effect* of casualty control can be described as a chain of events that occur in rapid succession, beginning with one casualty and ending in many. Each casualty triggers one or more additional casualties which, in turn, cause more casualties until the end result is completely out of proportion to the initial malfunction. Maximum effort must be extended to break the chain quickly and reestablish control of the situation. The following is an example of what can happen under a particular set of circumstances.

> Assume a combatant ship which has two engine compartments and two firerooms. The electrical plant consists of both a forward and after ship's service turbo-generator (SSTG) and switchboard. In addition, there is an emergency diesel generator and associated switchboard located forward. Assume that the steam and electrical systems are split fore and aft with two boilers on the line.
>
> The ship is steaming at standard speed outbound in a narrow ship channel. The bridge rings up all engines ahead full on the EOT. In attempted compliance with the OOD's order, the remote-reach rod in the after fireroom for one of the forced-draft blowers is broken, thus reducing boiler air supply. The after engineroom, unaware of the problem in the fireroom, opens the throttle too wide and drops the boiler pressure dangerously low. The main feedpumps cannot keep up with the sudden demand for more water, and a low-water casualty results. The after boiler is immediately secured and the engineer officer orders the steam plant cross-connected. Since the after fireroom personnel are busy securing their boiler, they are late in opening their assigned cross-connect valves. This slow action prevents steam from the forward plant reaching the after ship service turbo-generator (SSTG) and it drops off the line due to loss of steam

> pressure. When normal electrical power aft is lost the bus-ties are closed to allow the forward SSTG to feed the after switchboard. However, the electricians fail to strip the after board sufficiently and the forward SSTG trips off the line due to overload. When all normal electrical power is lost, the emergency diesel starts up automatically but fails to take the electrical load because a circuit breaker has stuck in the closed position. At this point the steering alarm sounds and the bridge reports that all steering control has been lost and the ship is heading into shoal water! In a desperate attempt to avoid going aground, the engines are backed full and the remaining boiler is dragged off the line by an overeager engineroom crew. Only minutes have elapsed between the time of the initial casualty and the moment when the ship crunches slowly to a halt, hard aground!

The above fictitious example is perhaps exaggerated, but it emphasizes the interdependence of engineering components and systems with the personnel who operate them. A highly developed ability to comprehend this inter-relationship and foresee the possible consequences is the hallmark of a professional shipboard engineer.

ENGINEERING OPERATIONAL SEQUENCING SYSTEM AND ENGINEERING OPERATIONAL CASUALTY CONTROL

The fact that there are many types of engineering plants in the fleet whose complexity requires in-depth operational knowledge by engineering personnel at all levels, is responsible for the evolution of the Engineering Operational Sequencing System (EOSS).

Basically, the EOSS is a written set of systematic and detailed procedures, in the form of charts, instruction cards, and diagrams, to provide the information required for operation of a shipboard propulsion plant. Although a very small percentage of the total operating time of any propulsion plant is spent operating under casualty conditions, that is the time when there is an urgent need for a detailed plan of the precise action that should be taken. The Engineering Operational Casualty Control (EOCC) has therefore been developed under the Operational Sequencing Program to fill this need.

The obvious objective of the EOCC system is to maintain vital engineering services at maximum capability whenever a casualty occurs and to prevent the spread of the casualty to other segments of the overall plant. Immediate response to a casualty also reduces the possibility of injury to personnel. Even during normal operating conditions the prevention of casualties must be the concern of all personnel, and this involves the proper training of all crew members.

The EOSS is currently installed on several aircraft carriers. Modified versions of the system are on several other ships.

As an enlargement of the Engineering Operating Procedures (EOP) developed to date, NavShips has developed an Engineering Operational Casualty Control (EOCC) manual which is provided as part of the system package. This manual provides efficient, technically correct procedures for casualty control and prevention pertaining to all phases of the engineering plant. The EOCC documents elaborate on casualties caused by personnel operating error, material failure, and battle damage, and are directly related to the operating procedures of the system. To date, EOCC manuals have been developed for a few ships, and they are being modified to reflect the comments from the ships that have used them.

The EOCC manual includes a casualty sequencing chart. In one column it lists a single-source casualty which, if not properly controlled, can cause the condition or casualty listed horizontally to its right in another column. Similarly, the items in the second column, if not properly controlled, can cause the damage listed in the third column. The idea is, of course, to give a quick reference to the action that should be taken to prevent the spread of casualties. Trouble-shooting charts, with listings of trouble, probable cause, and response to be taken, are also included. Trouble-shooting charts are provided for the malfunction of many mechanically operated components. The action taken by any watch stander can be critical in preventing the occurrence of a series of casualties which could easily lead to the loss of the entire propulsion system. A casualty sequencing chart is shown in Figure 8.

The format and phraseology of EOSS is standardized for all ships insofar as possible; in this regard, it is compatible with the Planned Maintenance System (PMS), and it can be adapted by personnel transferred from one ship to another or from one space to another. It does not eliminate the need for skilled plant operators; no program or system can do that. It is intended to provide the best possible utilization of the manpower and skills available in the crew. As this system is applied to more and more ships, more and more personnel will become involved, and every engineering officer should be aware of the existence of these new procedures. Its operation in the area of casualty control, in the form of a step-by-step outline of action to be taken, can be invaluable.

NavShips has underway a program to completely

THIS CHART IS USED FOR SEQUENCING CASUALTIES FOR MULTIPLE CASUALTY TRAINING. IF NOT PROPERLY HANDLED EACH SINGLE-SOURCE CASUALTY LISTED CAN CAUSE THE CASUALTY, PLANT CONDITION, OR RESPONSE IMMEDIATELY TO ITS RIGHT IN THE NEXT COLUMN.			
A. CAN CAUSE →	B. CAN CAUSE →	C. CAN CAUSE →	D.
1. Loss of automatic feed-water control.	1. High water in boiler (or water carryover)	Stop main engine. Stop turbogenerator.	
1. Loss of automatic feed-water control.	2. Low water in boiler.	1. Loss of steam pressure (above 1050 PSI). Slow main engine.	Loss of steam pressure (below 1050 PSI). X-connect.
2. Low water in deaerating feed tank.	2. Low water in boiler.	1. Loss of steam pressure (above 1050 PSI). Slow main engine.	Loss of steam pressure (below 1050 PSI). X-connect.
3. Ruptured deaerating feed tank or feed piping.	2. Low water in boiler.	1. Loss of steam pressure (above 1050 PSI). Slow main engine.	Loss of steam pressure (below 1050 PSI). X-connect.
4. High water in deaerating feed tank			
5. Loss of main feed booster pump.	2. Low water in boiler.	1. Loss of steam pressure (above 1050 PSI). Slow main engine.	Loss of steam pressure (below 1050 PSI). X-connect.
6. Loss of main feed pump.	2. Low water in boiler.	1. Loss of steam pressure (above 1050 PSI). Slow main engine.	Loss of steam pressure (below 1050 PSI). X-connect.
7. High salinity.			
8. Oil in fuel oil heater drains.			

Naval Engineers Journal, December 1970

Figure 8—Feedwater Casualty Sequencing Chart

develop, produce, and install the EOSS, including EOP and EOCC, on a wide variety of ship types. This includes installations on 15 different ship classes, involving ships of the CVA, DE, DEG, DD, DDG, DLG, LST, AFS, and LKA types. In all cases the ships under development are under 20 years of age.

SUMMARY

NavShips *Technical Manual,* Chapter 9880, Section III, contains many specific details regarding the concepts and methods of good engineering casualty control and a comprehensive discussion of electrical casualties. The following are some of the common causes of casualties, for which general recommended control procedures can be found.

FIREROOM CASUALTIES

Loss of fuel-oil suction
Failure of fuel-oil service pump
Loss of feed suction
Inability to build up or maintain feedwater pressure
Failure of emergency feedpump to take suction
Loss of feed pressure
Boiler tube or other pressure part carried away
Water glass carried away on boiler
Low water in boiler
High water in boiler—serious priming
Major fuel leak
Fuel oil in fuel heater drain (steam side)
Water in fuel oil
Fuel-oil pump air-bound
Forced-draft blower failure

Brickwork failure
Reduction of chloride (salinity) content in steaming boiler
Explosion in boiler casing while lighting-off
Fuel-oil fire
Flareback
Cold-boiler start
Loss of control air pressure
Loss of, or low, main feed booster pump pressure

ENGINEROOM CASUALTIES

Jammed throttle (ahead and astern)
Lubricating-oil cooler tube carried away
Loss of, or low, lubricating-oil pressure
Excessive lubricating-oil-pump discharge pressure
Lubrication oil leak into the engineroom
High oil level in the reduction gear case—oil emulsion
Locking and unlocking of shaft underway
Excessive vibration of shaft
Unusual noise in the reduction gear
Metallic noise emanating from turbine
Vibration of turbine
Hot bearings
Loss of vacuum in condenser
Leak in condenser
Empty feed bottom in use for make-up feed
Casualty to the deaerating feed tank
Drop in water level of deaerating feed tank during steady steaming
Deaerating feed tank too full
Breakage of evaporator gage glass
Loss of steam pressure in the engineroom

Cooling water to auxiliaries fails
Unusual noise from main feed pump when starting
Low, or loss of, main feed booster pump pressure

DIESEL ENGINE CASUALTIES

Sudden stopping of engine
Loss of air pressure to air clutch (LST type)
Electric power failure to governor control motor and air clutch
Inability to start engine (electric starting)
Low lubrication-oil pressure
Overspeeding of engine
Leak in lubrication-oil cooler
High fresh-water cooling temperature
Low sea-water cooling pressure
Failure of reduction gear cooling water and lubrication-oil pump (LST type)
Failure of reduction gear cooling water and lubrication-oil pump (AOG type)
Failure of speed-restricting governor
Loss of starting air
Loss of fuel-oil pressure
Water in engine cylinder and/or crankcase of air intake ports
Ship-service generators #1 and #2 disabled (AOG type)
Loss of sanitary water pressure

BATTLE DAMAGE CASUALTIES

Ruptured main steam piping (split plant)
Rupture in auxiliary steam piping
Rupture in auxiliary exhaust piping
Rupture in main feed piping

Rupture in fuel-oil suction and transfer piping
Rupture in high-pressure drain piping
Rupture in fuel-oil heating drain piping
Rupture in firemain piping
Rupture in sea-water cooling service piping
Main turbine casualty caused by shock of an explosion outside of the hull
Fireroom explosion–torpedo hit
Fireroom explosion–shell hit
Engineroom explosion–torpedo hit
Engineroom explosion–shell hit
Steering gear casualty–near-miss
Propeller casualty–near-miss

Every ship should have its own casualty control book written exclusively for its own particular plant arrangement. Type commanders frequently provide casualty control books for ships of the type they command, but it cannot be assumed that the procedures outlined therein apply to an individual ship and take into account any alterations that have been made. Information on the preparation of a casualty control book can be found in the latest NWIP-50 (series) *Battle Control.*

CHAPTER 8

Special Evolutions

COLD IRON

The term *cold iron* can be broadly defined as the condition of a ship's engineering plant when boilers are secured and vital services are being received from the pier. The number of vital services needed will vary, depending on the type of ship and the existing circumstances. For example, a destroyer undergoing a two-week restricted availability might receive all her basic requirements from the pier or a tender. Such major services would probably include:

electrical power;
steam for hotel services;
fire and flushing water.

Of course, there are other shore services that might reasonably be considered vital to a ship. Additional telephone lines, potable water, and fuel oil are certainly among them, but they fall into the category of routine services frequently provided during extended in-port periods. They are not indicators of, nor do they have any direct effect on, the material readiness of the engineering plant.

The frequency with which a ship goes cold iron is roughly in inverse proportion to her size. The bigger the ship the less often she will go cold iron. As a matter of fact, many large ships such as aircraft carriers go from overhaul to overhaul without ever going to this condition. There are three reasons for this:

1. The in-house capability of large ships to make repairs on idle machinery during in-port periods is much greater than is that of small ships.
2. Under normal circumstances, large ships have sufficient personnel to stand in-port engineering watches without working a hardship on the crew.
3. Most piers simply do not have the capacity to furnish the enormous quantities of services required by capital ships.

Whether large or small, the most critical vital service a ship receives when in port is shore power.

SHORE POWER

The term *shore power* commonly refers to any electrical power being received by a ship from an outside source whether it be a shore power plant or another ship, such as a destroyer tender (AD).

Most naval vessels are fitted with shore-power connections located at or near the main weather deck. Portable cables leading from the pier or from a ship alongside can be attached to these connections to supply power for the receiving ship's electrical distribution system.

It may be desirable to take on shore power for a variety of reasons such as: limiting the steaming hours on boilers, conserving fuel, minimizing in-port watch-standing requirements, or accomplishing maintenance and repairs on the ship's service turbo-generators and other machinery. Whatever the reasons, if certain minimum precautions are adhered to, taking on shore power can be a successful operation. On the other hand, poor planning, an indifferent attitude toward safety, and low standards of performance among cold iron watch standers can make it a totally disastrous exercise. Occurrences of shore-power-cable fires, explosions at connection boxes, and burned-up electrical or electronic equipment are not uncommon.

Planning Ahead

One of the most important facts to be remembered about shore power is that the power capacities of both the supplier and the user are usually limited. Therefore, a workable plan must be devised to limit the electrical load to a size that can be handled safely by the ship's shore-power capacity or the pier's shore-power capacity, whichever is the smaller. Keep in mind that during the hot months, when ventilating fans are working at full capacity and all air-conditioning machinery is in operation, a ship's electrical load is heavier than it is during the cold months. Secondly, permission to start heavy auxiliary or weapons machinery must be coordinated with a responsible person: high starting loads on heavy machinery are a common cause of overloading circuit breakers and causing them to trip off the power supply at the switchboard. Incidentally, this same precaution may be necessary during in-port periods when a ship is generating her own power with a reduced number of generators.

Before a ship goes on shore power, her shore-power cables, if carried, should be checked for broken insulation, and their terminals should be cleaned and tightened. Large ships do not normally carry their own shore-power cables and are at the mercy of the supplier. Therefore, the cables delivered should be inspected when they arrive at the berth. Loose connections on cable terminals are a constant hazard.

Strict observance of safety precautions is a must in handling shore power. One person should be designated "in charge" of the entire evolution. This will ensure a controlled sequence of events and make for better coordination of phasing, de-energizing sections of the ship's service switchboard, and re-energizing the system. Because of the desire to keep the nuclear reactor in operation, nuclear-powered ships continue to use some of their own power while paralleling with shore power, but it is still routine practice for conventionally powered ships to secure all internal electrical power before taking on shore power. Operating instructions vary with different installations: where none exist, the *Electrician's Mate 1 & C* (NavPers 10549 series) can be consulted.

Once a ship is on shore power, continuity of power should be maintained. The receiving ship has no control over the source of the power, but, by monitoring its flow, ship's personnel can do much to ensure uninterrupted service.

The Shore-Power Log

A shore-power log should be provided and a roving watch established to record periodically the bus voltage and the load readings. On ships not equipped with a permanently installed shore-power bus ammeter, a portable ammeter of the split-core type can be used. It should be understood that each installed ammeter registers only the current on the shore-power *bus* to which it is attached: on large ships there may be *four or more* shore-power cables feeding one bus. In most cases, each portable cable is served by a single circuit breaker on the pier or the tender. If one of these breakers trips, the remaining cables will take up the load and, should they become overloaded, electrical fires and explosions can result. Only by checking the individual phases of each cable *at the shore-power terminal box on the ship* can it be seen whether or not all cables are carrying approximately the same loads. If this is done the potential for malfunctions will be minimized.

In summary, there is no substitute for safety-consciousness and vigilance in the shore-power evolution. Monitoring the actual connecting and disconnecting of shore power requires close on-the-job supervision by experienced ship personnel. While shore power is being received, regular checks of the cable loads and scrupulous observance of all applicable safety precautions are absolute musts.

SHORE STEAM

While the ship's boilers are secured and the ship is utilizing shore power, most of the machinery in the engineering spaces is idle. The engineering spaces are termed *dead,* and the watches consist mainly of roving patrols. These patrols must keep a close check on the security of all unmanned spaces, and must be particularly alert for fire and flooding.

One of the most insidious and destructive hazards to a ship that is in the cold-iron condition for an appreciable length of time, is humidity. It can cause a multitude of problems with shipboard equipment:

1. electrical insulation on motor windings and controls absorbs the moisture readily, a tendency that increases as the insulation gets older.
2. rust forms on starter contactors and other metal parts.
3. pump motors and other electrically powered auxiliaries indicate lower and lower insulation to ground readings, as they are affected by moisture.
4. the inner races of ball bearings become pitted from corrosion.
5. the metal surfaces of solenoid-operated motor starters often rust so badly that the starters "hang up" at the next attempted use. (When this occurs the contact may either fail to close completely, thus burning up the starting coil, or it may close and stick in position making it necessary to open the feeder breaker in order to secure the unit.)

Perhaps the most frustrating part of humidity is that its full effect is not apparent until it is time to light-off the engineering plant in preparation for going back on ship's power. The time element of getting underway is often short and the problems begin to compound themselves. A condensate pump burns out or a gland-exhaust fan overloads. Then the in-port fuel-oil service pump won't start or one of the main ventilation blowers in the fireroom won't stay on the line. Most of these malfunctions can be prevented and many Casualty Summary Reports (CASREPs) need never be written, if a few basic rules are followed meticulously.

Hazards and Preventive Measures

One of the most common sources of moisture is shore steam being taken on for heating purposes. Leaking steam traps in the steam-heat drain lines can permit live steam to pass directly to the low-pressure drain tank in the bilges of a main machinery space. Huge clouds of steam billow forth and condense on the cool metal surfaces throughout the space. If this kind of situation goes uncorrected, trouble most certainly lies ahead. The solution is difficult and time-consuming—fix the steam traps. There are literally hundreds of steam traps installed in every steam-driven ship in the Navy, and the only way to keep ahead of their malfunction is conscientiously to inspect, repair, and replace them on a regular basis. A typical part of such a planned effort would be to renovate heating-system steam traps during the summer months when the system is idle.

Even when all steam traps are functioning perfectly, the engineering spaces, by their very nature, retain a highly humid atmosphere. Because of this, a primary effort must be extended toward keeping the machinery spaces and the equipment in them as clean and dry as possible. A simple waterproof cover placed over the electrical controllers and motors is a start. A safety-guarded incandescent lamp rigged near the windings of the larger motors that are located in especially humid areas can minimize the amount of moisture absorbed by the insulation material. Equipment that is in operating condition should be run from time to time to prevent moisture from accumulating in the electrical windings and to minimize the pitting of bearing surfaces.

Unfortunately, it is much easier to talk about taking these precautions than it is to take them. There is a natural tendency to follow such a program for a brief period of time and then to abandon it because it isn't considered worth the trouble. A large measure of common sense must be applied to arrive at a solution that is both logical and workable. The complete understanding and cooperation of all involved, including yard or tender personnel, is essential to the success of any such undertaking.

RUNNING THE DEGAUSSING RANGE

A steel-hulled ship should run a degaussing range for a magnetic check once a year, and more often if possible. Locations of operating degaussing stations can be found in the ship's Degaussing Folder. When it is determined that a magnetic check should be run, the appropriate degaussing station should be contacted by telephone or correspondence to schedule the proposed ranging.

The degaussing stations are equipped to measure and record the magnetic field of ships that pass over measuring equipment located at or near the bottom of the ship channel. A ship is said to be *ranged* when her magnetic field has been measured by this equipment. There are two types of ranging:

1. *Check Ranging.* This is the most common type of ranging, to which almost all ships are frequently subjected. It checks the performance of installed degaussing equipment and operating personnel.
2. *Calibration Ranging.* This type of ranging is used to determine coil settings and to derive information for the degaussing charts of newly constructed or modified ships. If performance during a routine check ranging is unsatisfactory, calibration ranging may be required.

After a time is scheduled for ranging, the degaussing settings for the ship should be adjusted for the local geographical area, in accordance with the charts in the Degaussing Folder. Voice communications with the Degaussing Control Station should be established sufficiently in advance to allow an orderly exchange of information before the ship enters the range. The ship must pass directly over the degaussing range at a constant speed set by Degaussing Control and based upon heading, length of the ship, and response time of recording instruments. Coil readings and the ship's fore and aft draft to the nearest six inches must be relayed by voice to the Degaussing Control. For a good check ranging, it is essential that, during transit, the ship remain at the set speed and on a course perpendicular to the installed range.

The results of the check ranging will be made known to the ship by correspondence. If they are unsatisfactory, a thorough investigation of the shipboard equipment should be made prior to re-running the check range. More explicit details can be found in NavShips *Technical Manual,* Chapter 9813.

TRANSITING FRESH WATER

There are certain hazards involved in a seagoing ship transiting or anchoring in fresh or brackish water, but they can be minimized, if anticipated. The major problem is caused by the marine life that clings to the inner walls of salt-water piping systems. After a few hours in fresh water, some of this marine growth flushes loose and is carried out through the various overboard discharges, which include flushing water from the heads and the outlets of various salt-water coolers and eductors throughout the ship. If operations permit, it is a good idea to lead fire hoses overboard from topside fire stations to discharge the firemain and permit the fresh water to purge the entire firemain system.

The effects of a fresh-water transit may begin after a few hours of exposure and may occur intermittently for a week or so. The age of the ship, her previous operating environment—marine life thrives in tropical waters—and the length of time since she made her last fresh-water transit, are all factors that influence the severity of the effects. One of the first things to expect is overheated machinery. It is quite possible for a ship's service turbo-generator to become overheated and have to be taken off the line because of clogged lube-oil coolers. Eductors may back up and cause overflows or flooding in heads and machinery spaces throughout the ship. There is really very little that can be done to prevent these occurrences, except to anticipate them and thereby reduce their deleterious effects.

When long periods are to be spent at anchor or moored in brackish or fresh water, the evaporators must be adjusted to compensate for the difference in the chemical composition of the water.

REFRESHER TRAINING

The subject of refresher training in ships of the fleet is greeted with no enthusiasm by the crew. It is unpopular in the fullest meaning of the word. Any engineer who has ever been through refresher training remembers the long hot days spent in the mass confusion of drills, drills, drills—and the long hot nights spent trying to find out what was done wrong during the long hot day. In this kind of atmosphere it is all important to keep the task at hand in realistic perspective.

Too often the engineers, or the entire ship's company, start off on the wrong foot and never recover. The Fleet Training Group instructors are frequently looked upon as unnecessary people who interrupt and complicate life aboard ship. It is important to consider that their job may be even tougher than the engineer officer's. Month after month they see ships come and go, and they watch people make the same mistakes over and over again. It is little wonder that they lose patience occasionally. Most of the men in Fleet Training Groups are highly qualified, truly dedicated professionals and they are there to help a ship achieve a reasonable degree of battle readiness. They, too, take pride in a job well done.

Much has been said about the "can do" spirit that prevails in the Navy. There are advantages and disadvantages to such a spirit as it applies to varied situations, but in refresher training it is the key to success. Without it, enthusiasm lags and fatigue, both mental and physical, takes over. This is apt to be followed by mistakes, ruffled tempers, and a sense of futility which may contribute to ultimate failure.

No attempt will be made here to go into the details of refresher training. The "nine-pound package" of instructions will be received from the Fleet Training Group well in advance of the ship's reporting for training. Information relating to what must be installed, what forms must be filled out, etc. is all there —to be read and acted upon. There are a considerable number of forms to be completed and certain pieces of equipment have to be opened, and the majority of it should be accomplished prior to arrival at the training point. It is imperative that advance instructions be carried out to the letter. If they state that all hatches, doors, and scuttles must have identification markings on *both sides* of the access, the crew should turn to and mark them on *both sides.* If these instructions are not followed, many long hours during the training period will be wasted, as frantic efforts to comply are made. There is also a distinct psychological advantage to be gained from getting things done ahead of time. When a ship arrives with all the preliminary planning and work completed, the Fleet Training Group's first impression is favorable. They know that an effort has been made and that all parties will enter into the operation in a better frame of mind. No one expects training to start in a state of perfect readiness: however, it is expected that the ship will be *ready for training*. This makes the task of refresher training measurably easier.

Don't be discouraged if the first few days produce

unsatisfactory grades across the board—they almost always do; it's part of the ritual. Keep plugging away, emphasizing to the crew the learning aspect of refresher training. This may be one of the few opportunities an engineer will have to really dig into the business of casualty control and damage control on a large scale. It can produce a battle-ready ship or a battle-weary crew, depending on how it is approached. Enthusiasm engenders a spirit of competition; it also relieves the pressure. In effect, refresher training is like having a private war—and it must be won.

CHAPTER 9

Fuel and Water

The topics discussed in this chapter apply mainly to the fuel and water requirements of a conventional steam power plant with a pressure-closed feed system.

FUEL AND WATER LABORATORY

FUNCTION AND IMPORTANCE

A man driving a car likes to know that his fuel gage is accurate—and so it is with the captain of a ship. The vital task of keeping a constant check on both the quantity and quality of feedwater and fuel on board ship falls to the men of the Fuel and Water Lab.

Physically, the Fuel and Water Lab may be a small cabinet of test equipment mounted in one corner of the log room, or it may be a spacious compartment designed as a testing laboratory and fueling station. Similarly, the oil king and his permanently assigned crew may range from one or two men on a destroyer, to a group of twenty or more on an attack carrier. Whether it is a one-man operation or a large organization, its task is clear.

From receipt to expenditure, every gallon of water and fuel oil on board must be accounted for. It must be accurately inventoried, properly stored, efficiently transferred, and continuously tested for purity. This is a never-ending, year-round task that requires meticulous attention to detail. One faulty feedwater test, one inaccurate tank sounding, or one brief moment of inattention can result in anything from minor inconvenience to total catastrophe. The importance of a smoothly functioning Fuel and Water Lab is recognized by every experienced engineer.

CHOOSING AN OIL KING

Since the oil king occupies a key billet in the Fuel and Water Lab as well as in the Engineering Department, the choosing of the right man for the job is of major importance. Some large ships solve this problem by simply designating the senior boiler technician in the department to be the oil king.

Regardless of the method used to choose the oil king, keep in mind that he must be thoroughly familiar with the ship. He should be an authority on the valve arrangements, piping systems, fueling procedures, and water testing and treatment methods on his ship. He should provide the kind of leadership that inspires alert, intelligent response from the people with whom he works. Whether it be an emergency refueling at sea in the middle of the night, or a dull in-port watch on a Sunday afternoon, the men who monitor and control the ship's water and fuel oil must perform reliably.

RECORDS AND REPORTS

It is absolutely mandatory that the Fuel and Water Lab have an up-to-the-minute and accurate system of record-keeping and reporting, and that every man who works with fuel and water clearly understand how it works and why it is necessary. This system should cover every aspect of report-making and record-keeping so that it may accommodate permanent information such as Boiler Record Sheets, and transient data such as tank soundings. For example, fuel and water consumption plotted in graphical form on an hourly or watch-to-watch basis can be invaluable in establishing usage trends and spotting potential trouble. Consider for a moment some of the immediate and long-range consequences of not having an up-to-date and accurate system:

1. If the record of water-produced versus water-used is not accurate
   a. the ship could run dangerously low on water —suddenly and unexpectedly.
   b. weeks or months could elapse with the engineer officer *mistakenly* thinking the plant is in excellent condition—or vice versa—because feedwater consumption is the major criterion for judging plant performance.
   c. the crew could be wrongly accused of using potable water that was never produced.
2. If fuel-tank sounding and testing procedures are not properly conducted
   a. oil spills, large in size and frequent in number, could occur.

   b. service tanks could run dry at a critical moment, causing a chain reaction of casualties that could place the ship dead in the water, aground, or in a collision.
   c. water could get into the fuel oil, and the immediate effects could be the same as in b. above. The long-range effect could be boiler firesides damage.
   d. the inventory of burnable fuel remaining on board could be inaccurate, which could cause the ship to run dangerously low on fuel.
3. If the supply records are not accurate, the ship could get underway for an extended period and be missing or run out of
   a. boiler compound
   b. water-testing chemicals
   c. chloride disinfectant for potable water
   d. sounding tapes, rags, etc.

The above lists are far from complete but they serve to point up the fact that shipboard engineering, like any complex operation, covers a wide range of large and small details, which must be identified and dealt with intelligently if a reasonable measure of success is to be achieved. This can be accomplished only through a timely and accurate system of record-keeping and reporting.

CARE AND HANDLING OF FUELS

IDENTIFYING THE PROBLEM

Nowhere is the need for expert performance more evident than in the functions that involve the care and handling of fuels. A whole range of petroleum products, from high-test aviation gasoline to the familiar Navy Special Fuel Oil (NSFO), is carried on board Navy ships. The potential for fire, contamination, and spillage is ever-present and must be guarded against constantly. The most challenging problems in this area occur on aircraft carriers where planes must be fueled with highly volatile liquids in the presence of pure oxygen, high explosives, and static electricity. It takes the coordinated efforts of many well-trained officers and men to control safely the numerous fueling operations so fundamental to successful carrier operations.

Most ships are, however, faced with much simpler problems than those of the carrier. For example, in routine circumstances, a typical destroyer would carry only two types of fuel:

1. NSFO, for main propulsion
2. Diesel fuel, for emergency diesel generators and ship's boats.

Whether fueling demands are simple or complicated, three considerations are essential in the care and handling of fuels:

1. maintenance of the required standards of purity
2. strict control over its movement
3. enforcement of applicable safety precautions.

Concentrating on NSFO, since it will be the cause of most problems, consider the following:

PURITY

The quality of NSFO is reasonably high. From shore base to tanker, to consuming ship, it is fairly free of contamination. When contamination does occur, it most commonly takes the form of salt water that enters through leaky tanks or is present in the remnants of ballast water. Salt water in the oil causes two specific casualties in the boiler:

1. loss of fires
2. heavy build-up of slag on firesides of tubes and firebrick walls.

NavShips *Technical Manual,* Chapter 9550, lists procedures for testing fuel oil and the requirements for taking a sample while refueling is underway. It should be noted that the quantity and quality of oil received is the responsibility of the ship *receiving* fuel.

The above is not intended to create the impression that fuel oil contamination occurs only during refuelings. It is entirely possible for a perfectly good load of oil to become contaminated in the ship's tanks. To quote NavShips *Technical Manual,* Chapter 9550:

> Oil being carried in the tanks of consuming ships should be tested for water contamination once each week and before being drawn upon for use.

This is a sometimes difficult, but very good, rule to follow. The important idea that should be stressed is this: the ultimate objective of the entire effort is to find out whether or not there is contaminated oil on board the ship. The principal danger arises when the contamination goes *undetected.* Once contamination is known to exist, steps can be taken to allow tanks to settle out for stripping. Extreme caution should always be taken to ensure that the fuel-oil service tanks are kept absolutely free of water. Some engineers, aware of the need for this precaution, may hesitate to use fuel stowage tanks for ballast, and thereby at times could critically affect the stability posture of their ship.

MOVING FUEL

General

There are three major fuel-moving occasions:

1. When it is being received
2. When it is being delivered
3. When it is being transferred from one part of a ship to another.

Since only large steam-driven ships have the capacity to deliver fuel, this discussion will be confined to receipt and internal transfer. The fundamental rules to follow when fuel is being moved are simple and easy to remember: know the fuel system; be painstaking and thorough; and establish good internal communications between sounding crews and topside receiving-station personnel.

There is no substitute for knowledge. This is as true in moving fuel as it is in conning a ship. Every man involved should not only be familiar with the part of the fuel system he has to work with, but he should know how it fits into the overall scheme of things. About the only effective way to accomplish this is through system tracing. An unpleasant task it may be—but it pays off.

In addition to knowing the layout of the fuel system, the officers and men in the Engineering Department should thoroughly understand fueling procedures and what their own particular jobs consist of. This may sound like a ridiculous statement—but here are a few examples of incidents that happen every day in the fleet.

a. A young officer is sent to the topside refueling station on the spur of the moment to act as officer-in-charge. He arrives at his station, slaps on his life jacket, then—because he has absolutely no idea of what is going on—he stands quietly aside. Terms such as *blowdown* and *back suction* are merely colorful expressions that have no meaning to him. His only opportunity to get into the act will occur when something goes wrong and he suddenly finds himself held responsible.

b. During a nighttime refueling, a fireman who has just reported aboard is hurriedly sent back to the after crew's living compartment to sound a tank. By the time he finds a place to plug in his headphones and locates the sounding tube (under the bottom bunk in the after starboard corner of the compartment), the tank has overflowed.

c. An inexperienced man is sent to line up the fuel system. He leaves valves open that should be closed and valves closed that should be open. No one bothers to check the line-up after him. The result—a major oil spill.

If the moving of fuel is planned, there is no problem in finding the extra time needed to be painstaking and thorough. Often, the fuel system can be partially lined-up before the taking-on of fuel begins, leaving only a few key valves to be positioned and checked. It should be standard policy to check every valve position prior to moving fuel. When the fueling detail is set, the stations should be manned expeditiously. A word of philosophy is appropriate here. Too often, an over-anxious officer of the deck will station the refueling detail much too far in advance. In any shipboard evolution, constant false alarms and long hours of standing by on station tend to lower morale and impair performance. Certainly, some long waits cannot be avoided—but they should not become a matter of routine.

Once fueling stations have been manned, communications should be checked out. It's a good idea to have a few extra sound-powered phone sets immediately available in case of a bad set on station.

The oil king, or whoever mans the phones at the fueling control station, should have complete control of the line. Control of the positioning of valves, the lighting-off and securing of pumps, the sounding of tanks, and other details, must be *absolute* and must be exercised through sound-powered communications. The degree of control necessary cannot be accomplished without strict communications discipline.

With proper training and a little practice the business of refueling at sea can become an efficient routine operation. However, there is still one remaining aspect of moving fuel which deserves special attention.

In Port

> Except as authorized by law or by regulations issued by competent authority, no refuse or oil shall be discharged into U. S. inland or coastal navigable waters, nor shall oil be discharged within any prohibited zone specified in the Oil Pollution Act, 1961, and any

> amendments thereto. [*Navy Regulations,* 1948, Article 1272.]

The Oil Pollution Act of 1924 (as amended) and the Oil Pollution Act of 1961 are both in effect. The 1961 act broadens and extends the 1924 act.

The legal, moral, and diplomatic implications of discharging oil into the navigable coastal waters of any country are quite clear. It is the responsibility of every engineer involved in the movement of fuel to exercise extreme caution when moving oil in port or in close proximity to land. From a practical point of view, much can be done *prior to entering port* to reduce the chance of oil spillage. Here are a few procedures which, if accomplished while still at sea, can simplify the problem considerably:

1. pump all bilges
2. strip dry all contaminated tanks
3. complete all such internal transfers of oil as
   a. pumping up all service tanks
   b. shifting oil to trim the ship
   c. shifting oil preparatory to refueling in port.

When fuel must be moved in port, it should be done with the utmost care. Oil spills often result from the most routine evolutions, which seem so simple that short cuts are taken and errors are made. For refueling in port:

(1) When scheduling the refueling, be sure that it does not conflict with another event, such as handling ammunition.
(2) Have a standard organization.
(3) Move no oil until the fuel system has been lined

up properly and checked by someone other than the person detailed to do the job.

(4) Move no oil until good communications have been established with *all stations.*

(5) Station lookouts topside to watch for oil spills, and be sure that they are on the fueling detail's sound-powered phone circuit.

(6) Before refueling begins, find out the local procedures to be followed and the telephone numbers to be called in case of an oil spill. *Time* can be of primary importance in halting the spread of an oil spill.

Modifications to the above procedures might reasonably be made for minor transfers of oil, such as pumping up a service tank, but the potential for spillage always exists and must be guarded against.

SAFETY PRECAUTIONS

Safety precautions applying to "Hot work in way of flammable or explosive materials and entry into closed or poorly ventilated spaces" are covered in great detail in the NavShips *Technical Manual,* Chapter 9920. They should be studied carefully. In addition, Chapter 9550 lays down specific precautions to be taken while fuel oil is being received or discharged:

> a. The use of naked lights, electrical or mechanical apparatus likely to spark shall not be permitted within 50 feet of an oil hose in use, an open fuel tank, the vent terminus from a fuel tank, or an area where fuel oil or fuel oil vapors are present or may reasonably be expected to be present during fueling. . . .
>
> b. Portable electric lights used during fueling shall

have explosion proof protected globes and shall be thoroughly inspected for proper insulation, and tested prior to being used.

c. On board vessels being fueled, portholes in the ships structure on the side from which fuel is being received shall be closed and secured.

The precautions set forth above are an essential minimum for the exercise of safety while receiving or discharging fuel oil. The commanding officer shall issue such additional orders as his judgment dictates in respect to further limitations on the privilege of smoking or the use of equipment capable of producing spark or flame during the transfer of fuel oil.

THE NAVY DISTILLATE FUEL CONVERSION PROGRAM

Revision B, 10 December 1969, to The Navy Distillate Fuel (ND) Conversion Program Acquisition Master Plan states:

> The Navy is currently pursuing a far ranging program of converting all conventionally boilered surface ships to the capability of burning distillate fuel without eliminating the capability to burn residual (NSFO) fuel. The overall objectives of the program are:
>
> 1) To convert all in-service and new construction petroleum fuel steam-driven surface ships from Navy Special Fuel Oil to Navy Distillate Fuel.
>
> 2) To convert all Navy ship fuel oil consumption to a Navy Distillate Fuel as contrasted to maintaining a three-fuel surface Navy currently using Navy Special Fuel Oil, Diesel Fuel, and JP-5.
>
> Historically the United States Navy has burned residual fuel for steam generation in conventional surface ship propulsion systems and in associated in-port or "hotel" support systems. The principal favorable factors associated with the use of Navy Special Fuel Oil are essentially, worldwide availability and low unit cost.
>
> The principal adverse factors associated with the use of NSFO are:

1) Fouling of firesides of boilers by permissible impurities in Navy Special Fuel Oil, principally sulphur, ash, and carbon residue.

2) Decrease in ship readiness associated principally with cleaning firesides.

3) High corrosion rate of above-deck equipment associated with exposure to products of combustion of Navy Special Fuel Oil.

4) Substantially below average retention rate of Navy enlisted personnel who perform boiler cleaning operations.

There can be no argument with the fact that a one-fuel Navy using a clean-burning, high quality product would solve many shipboard problems. All aspects of handling, storing, and consuming fuel on board ship would be greatly simplified.

CARE AND HANDLING OF WATER

IDENTIFYING THE PROBLEM

The ability of a steam-driven ship to meet her operational commitments is directly bound to her ability to continuously produce a sufficient quantity of high-quality fresh water. When it becomes routine for the daily water consumption to exceed daily production, it is clear that the ship must initiate restrictive measures, or she will eventually run out of water. Normally, the crew are the first to be affected by restrictive measures in the form of "water hours." Water hours allow potable-water production to be temporarily diverted to feedwater production for use in the steam plant. If the steam plant's feedwater consumption cannot be reduced and evaporator production cannot be increased, the ship will eventually have to return to port—an operational casualty.

WATER PRODUCTION AND CONSUMPTION

Water supply on board ship can become a critical problem for several clearly identifiable reasons:

1. As ship grows older, fresh-water leaks accumulate and sometimes they are not repaired because
   a. funds are lacking
   b. time or opportunity is lacking
   c. they are not detected
   d. there is not a good program for detecting and repairing them.
2. Some ships have been converted or altered in such a way that fresh-water *demand* has been increased and evaporator capacity has not.
3. The evaporator maintenance program is poor, with the result that the evaporators operate at reduced capacity.
4. Evaporator operators are poorly trained or negligent, and that can lead to a great reduction in both the quantity and quality of fresh water produced.
5. The crew has not been sufficiently indoctrinated in the judicious use of potable water.
6. Maintenance of today's high standards of boiler-water purity necessitates an increased number of boiler blowdowns and dumps.

Whatever the problems may be, one thing is crystal clear: they must be solved if the ship is to operate with any kind of engineering dependability.

CORRECTIVE MEASURES

One of the first steps which must be taken to correct a water problem is to make an accurate estimate of the situation. It does no good to go dashing about the ship looking for leaks when the problem lies in poor evaporator performance. Here are a few thumb rules:

1. Be sure the water production and consumption figures are accurate. *Do not* rely on meter readings—use tank soundings over several time intervals.
2. Check the size of overload orifices, if such are installed on the evaporators.
3. Look for out-of-the-way areas of water consumption: garbage grinders that use fresh water for flushing are a good place to start.
4. Make sure all thermometers, pressure gages, salinity indicators, etc., installed on the evaporators are calibrated and accurate.
5. Institute a program of tagging all known leaks. If this is not done, when the ship pulls into port and the plant is secured, the leaks stop and are forgotten until the next underway period.
6. Embark on a program to make the engineers and the rest of the crew water-conscious. Don't take it for granted that every sailor knows how to take a Navy shower—many don't.
7. If an evaporator breaks down, start repairs *immediately*. Every hour it is off the line puts you that much further in the hole.

The road to reduced water consumption and increased water production can be a troubled one. On ships that operate under a daily threat of water hours, only a continuing effort, steadily applied, can bring measurable progress. A good engineer takes a great deal of pride in having a tight steam plant, as well he should, because it takes a lot of know-how and plenty of hard work to achieve and maintain one.

PURITY

The amount of boiler waterside maintenance that is required and the length of the useful life of the boiler itself, ultimately depend on the maintenance of high standards of boiler water and feedwater purity and treatment. Appendix C provides a summary of recommended feedwater tests.

NavShips *Technical Manual,* Chapter 9560, provides complete coverage of water treatment, purity, and the effects of various impurities. Some type commanders have established certain procedures to be followed and standards to be met, thereby indicating their concern in this vital area of shipboard engineering. For example, a type commander may require that every man who performs chemical tests on boiler water must be a graduate of one of the schools for boiler-water treatment. If a ship has no such requirement, it's a good idea to institute one.

Perhaps the most difficult problem associated with boiler-water treatment and purity is detection. Erroneous salinity-indicator readings, deteriorated water-testing chemicals, improper testing procedures, lack of necessary equipment, lack of knowledge and experience on the part of those charged with testing and treating water—all of these are obstacles to be overcome.

Technical competence and a conscientious attitude toward maintaining proper boiler-water standards must be developed if the ship's main propulsion boilers are to operate safely and efficiently.

CHAPTER 10

Trials and Exercises

TRIALS

Trials and exercises, like the inspections discussed in Chapter 3, are conducted to determine the ship's readiness for service, the extent of necessary repairs, the adequacy of completed repairs, and the acceptability of new construction and converted ships. Information relating to shipyard trials during overhaul are discussed in Chapter 12. This chapter deals with trials and exercises required of ships during the normal operational cycle.

SHIP-INSTITUTED TRIALS

The two most important engineering trials conducted periodically, as directed by the type commander, are the full-power trial and the fuel-economy trial.

Full-Power Trials

These trials have observers assigned, unless the ship is steaming independently under conditions where it is considered impractical to obtain outside observers.

The observers are usually from an afloat staff or are members of the Engineering Department of a similar class ship. In this manner, a truly impartial trial can be conducted and grading is not influenced by personal desires to make a good showing.

A day or two before that for which a full-power trial is scheduled, machinery and material should be inspected to ensure that the ship can run the test without danger to equipment or personnel. This inspection should cover boilers, main engines, pumps, safety devices, piping systems, and auxiliary machinery.

The engineer officer should report the readiness of the plant to the commanding officer no later than one day before the scheduled trial and he should state his opinion as to whether or not the plant is in proper condition to conduct the trial safely.

Before the trial begins, the engineer officer should

1. organize a system by which all fuel-oil and feed-water soundings can be read simultaneously.
2. provide the chief observer with a written statement as to the date of last undocking and the authorized and actual settings of all main machinery devices and dates of calibration.
3. be sure that draft, trim, and liquid loading conform to trial requirements.
4. obtain the required full-power rpm derived from calculations made by the chief observer based upon the ship's draft, trim, and liquid load at the time of trial, as directed by the type commander.

Just prior to or during the trial, the engineer officer should ensure that the following requirements are fulfilled

1. synchronize all ship clocks before the trial.
2. commence the trial on date set, provided it is light enough to witness smoke-prevention run (if required).
3. take and record fuel-oil, rpm, speed, and equipment-operating readings at least every half-hour.
4. determine fuel expenditures for each hourly interval.
5. maintain appropriate ship "material condition" during the trial.
6. provide normal ship services during the trial.

Scheduled trials should not be cancelled unless weather conditions are such that the ship might be damaged at high speeds, there is some serious material casualty, or there are situations that would endanger human life.

Prior to starting the full-power trial, the plant should be thoroughly warmed up and the speed of the engines gradually increased to the speed specified for the trial. This will not only help prevent casualties but will also provide better thermodynamic conditions, and, consequently, better results. Trials should not be conducted in shallow water. Data should be recorded frequently enough to allow a reasonably correct indication of the power developed to be obtained. Details of trial observation can be found in *Fleet Exercise Publication* (FXP-3).

Fuel-Economy Trials

These trials are prepared for and conducted in a manner quite similar to the full-power trials. Except for the full-power rpm consideration, all the items listed above pertain to the economy trials. Full details of the trial observation can be found in FXP-3.

TRIALS BY HIGHER AUTHORITY

Board of Inspection and Survey (Insurv)

MATERIAL INSPECTIONS (MI)

These inspections are carried out in accordance with the statutory requirements of Congress at certain stated times:

1. when a ship returns from permanent duty on a foreign station
2. when the fitness of a ship for further service is in doubt
3. when a ship has been declared unfit and is to be stricken from the Naval Vessel Register
4. at least every three years (if practicable)

Proper preparedness includes ensuring that all details requested by the INSURV Board are complied with, including preparation of INSURV Board Condition Sheets. Repair work should be kept to a minimum. The INSURV Board should be provided with the following information and records:

a. what tests were incomplete or unsatisfactory
b. what repairs are required
c. what equipment fails to meet performance
d. what alterations are desired

e. what field changes or shipalts have not been accomplished
f. what safety hazards there are
g. what installations are only partially completed
h. what items have not been completed since last INSURV inspection
i. Booklet of General Plans
j. Ship Information Book
k. Damage Control Book
l. Allowance List (COSAL)
m. Watertight Integrity Test Book
n. Docking Report
o. Engineering Logs
p. List of Approved Alterations
q. PMS/MDC data

PRELIMINARY ACCEPTANCE TRIALS (PAT)

These trials are conducted to determine whether or not a ship has been constructed or converted in accordance with the contract specifications. Prior to such a trial, the prospective engineer officer should provide the prospective commanding officer with a list of items he considers incomplete or unsatisfactory and a list of the alterations or improvements he considers essential to the mission of the ship. The PAT might include the following:

1. full-power trial
2. quick reversals and backing
3. anchor test
4. steering test
5. locked-shaft test
6. boiler overload test

FINAL ACCEPTANCE TRIALS (FAT)

Normally conducted six months after the PAT, these trials determine any failures of equipment and material caused by contractor error. In preparation for the trials a list of work required by the contract and specifications should be compiled, and the following reports, test data, and publications should be provided to the Board:

1. Waivers on uncompleted work
2. Damage Control Book
3. Ship Information Book
4. NavShips *Technical Manual*
5. Watertight Integrity Test Book
6. Test memoranda
7. Test Agenda
8. Docking Plan
9. Docking Report
10. Damage Control Diagrams

TRIALS UNDER NAVSHIPS COGNIZANCE

Builders' Trials

These trials are conducted by the contractor to demonstrate satisfactory ship operation. Dock trials are conducted to demonstrate to the supervisor of shipbuilding (SUPSHIP) and the prospective engineer officer of the ship that the engineering plant is ready for sea trials. Sea trials are conducted to demonstrate that the ship is seaworthy and ready for PAT. Prior to the sea trials the contractor and SUPSHIP should ensure that

1. all temporary rigging has been removed

2. bilges are clean and bilge eductor and pump systems work
3. all instrumentation has been certified
4. the magnetic compass has been adjusted
5. operational tests of ship systems have been conducted
6. ventilation and air-conditioning systems operate satisfactorily.

Special Trials

STANDARDIZATION TRIALS

NavShips selects one ship from each class of ships, either new construction or major conversion, to determine class standardization curves for various ship speeds, rpm, shaft horsepower, and shaft thrust, at both heavy and light displacement. Conditions for standardization trials are prescribed in NavShips *Technical Manual,* Chapter 9080, and in the trial agenda provided.

TACTICAL TRIALS

These trials consist of operational procedures to determine the characteristics of a ship relative to normal turning cycles, acceleration and deceleration in straight path and maneuvering, and special turns.

PLANT EFFICIENCY TRIALS

This type of trial determines the efficiency of the propulsion plant installations and tests heat balance, water rate, and fuel rate.

NOISE TRIALS

The purpose of noise trials is to determine

1. airborne noise
2. underway radiated noise
3. self noise
4. overside noise radiating underwater from individual equipment installed in the ship
5. structure-borne noise

VIBRATION TRIALS

These trials may involve one or all of the following tests:

1. hull vibration
2. propulsion-system vibration
3. vibration in other individual components, systems, and/or structures.

OTHER TRIALS AND INSPECTIONS

Pre-inactivation or Pre-overhaul Trials

Both these types of tests are conducted to determine the extent of ship repairs required, so that cost estimates may be prepared. The trials are conducted by the commanding officer. The engineer officer must be sure that a complete analysis of the condition of his machinery and needed repairs or alterations are properly entered in ship maintenance records.

Inactivation Inspections

The purpose of these inspections is to ensure that all required inactivation procedures have been completed or that exceptions have been noted, and required inventories are complete. They also determine whether the ship will or will not require docking.

Recommissioning Trials

A ship being returned to active status after having been inactivated must undergo recommissioning trials. These trials are usually conducted by the commanding officer after recommissioning, and include a dock trial. Their purpose is

1. to evaluate maintenance procedures during inactivation
2. to determine whether full-power requirements can be met
3. to test adequacy of repairs and alterations made during activation
4. to evaluate methods of preservation that were used
5. to familiarize the crew with full-power operating procedures.

EXERCISES

The U. S. Navy has published a series of Ship Exercises which are contained in FXP-3. These exercises are provided to evaluate the training of ship personnel as supervised and controlled by type commanders. They are usually conducted as part of the overall ship fleet competition. They do, however, provide a basis for ship training at the department level.

The engineer officer who desires to have his department excel in fleet exercise competition will not achieve his goal without considerable training. He must train his crew to handle these fleet exercise drills on a daily basis, since they represent situations from which the majority of actual casualties will occur on equipment under his cognizance. How does one prepare the crew for the conducting of such exercises? Certainly not by sneaking into the engineroom and tripping out a generator. With crew fluctuation as it is today, the following is considered to be a proper method of training a crew for engineering and damage-control exercises:

1. State verbally the drill to be conducted and then have personnel walk through their actions in that simulated drill.
2. When "simulated" confidence has been established, attempt minor drills by actually securing equipment and tell crew exactly what casualty has been conducted.
3. After detailed training, conduct actual casualties by notifying only the top watch what is taking place.

No exercise should be conducted without the prior knowledge and approval of the commanding officer. Furthermore, permission must be obtained from the OOD before conducting any exercise that will actually place equipment out of commission and could affect the ship's steaming condition.

SUMMARY

It must always be remembered that no matter how pressing daily routine becomes, an essential factor in maintaining a ship's battle readiness is repetitive training in established shipboard exercises. This training effort coupled to proper preparation for and satisfactory completion of shipboard trials will assist ships' officers in determining and improving the level of ship battle efficiency.

CHAPTER 11

Shipboard Maintenance and Repair

Maintenance and repair, as defined by the office of the Secretary of Defense, is the function of retaining material in, or restoring it to, a serviceable condition. This function includes modernizing, overhauling, rebuilding, reclaiming, servicing, and repairing.

Information on every phase of shipboard engineering is available aboard ship. It exists in a multitude of manufacturers' instruction manuals, Navy directives, pamphlets, bulletins, letters, and type commanders' instructions. In addition, detailed texts have been prepared by the Department of the Navy to outline the specific important phases of operation, care, and maintenance of shipboard machinery. A list of recommended reading is provided at the back of this Guide. This chapter will also discuss the Planned Maintenance System and Maintenance Data Collection (PMS/MDC) procedures and analyze the steps to be followed when placing a piece of equipment out of commission.

PROCEDURES FOR PLACING EQUIPMENT OUT OF COMMISSION

As a general rule, no equipment should be placed out of commission without the permission of the commanding officer. Needless to say, it is essential that he know exactly what equipment is out of commission so that he knows the ship's state of readiness.

Some equipment casualties and the placing of certain equipment out of commission for repairs, must be reported to higher levels of command in the form of a Casualty Report (CASREP). There are a few commanding officers who neglect to report such situations because they think a CASREP indicates poor performance on their part. This is the farthest thing from the truth. No piece of equipment with mechanical moving parts is immune from sustaining a malfunction. It is essential for superiors, such as type commanders, to know the condition of ships in their force. Think of the embarrassment if a ship were called upon by her type commander to perform an emergency duty, and was unable to get underway because of material conditions of which the type commander was not aware. At that point, the ship command would be in much more trouble than if the true material condition had been intelligently reported.

Guidance here is quite simple. Be sure that engineering personnel

1. keep well informed on the actual material condition of equipment under their cognizance.
2. make conditions known at "eight o'clock" reports.

3. where necessary, fill out CASREPs in accordance with current directives.

In placing a piece of equipment out of commission, it is necessary not only to analyze what effect this will have on the ship's readiness, but also what other equipment will be directly affected and what safety precautions should be taken to prevent an accident. For instance, if the watersides of a condenser are open and the sea valves are not wired shut, there exists a potential danger of flooding a machinery space. If a steam valve or a steam auxiliary turbine has been removed without steamline valves being locked shut, and someone energizes the steam system, electrical installations in the machinery space could be destroyed and personnel could be burned. Safety of the ship and of ship personnel must be considered at all times.

When the ship is operating, naval regulations are quite specific as to the requirement for reporting equipment which is out of commission to the officer of the deck and to the engineering officer of the watch. These requirements are set forth in *Navy Regulations,* 1948, Chapter 10, Articles 1028 and 1029, and are quoted in Chapter 6 of this Guide.

As a general rule, the engineering officer, when making his request to place equipment out of commission, should be able to provide answers to the following questions:

a. What is the equipment and what is its function?
b. What effect will placing it out of commission have on the operating capabilities of the ship?
c. Are repair parts aboard?
d. What is the estimated downtime?

e. What are the alternative solutions, if any?
f. What is his personal recommendation? (This recommendation should be made in a positive manner.)

After discussing the matter with the commanding officer, carry out his final orders. Under normal situations, if the request has been presented properly, the commanding officer will grant permission for placing equipment out of commission. Remember that he has the sole responsibility for the safety and readiness of the ship. He can delegate authority for control of the engineering plant to the engineer officer, but the engineer officer owes an accountability to the commanding officer at all times.

Once permission has been granted, all hands in the Engineering Department should be made aware of what equipment is to be placed out of commission. This should be done verbally at quarters and indicated on the master diagram at main engine control. Entries should also be made in applicable engineering logs specifying what equipment has been placed out of commission and what safety precautions have been taken. Finally, appropriate 3-M data should be submitted.

STANDARD NAVY MAINTENANCE AND MATERIAL MANAGEMENT SYSTEM (3-M SYSTEM)

An effective fleet maintenance program depends on several fundamental elements: facilities, funds, personnel, materials, and management. Prior to implementation of the 3-M system, shipboard maintenance programs were not standardized. They functioned only where there were interested, experienced personnel assigned to billets of responsibility, and they varied from ship to ship in method, technique, and results.

It is important to again take note of that old adage, "there are no shipyards at sea." Every ship must maintain herself to the best of her ability, and the degree of her ability depends upon how well the crew is trained to handle routine maintenance. Therefore, on-the-job training is essential. If repair-ship and shipyard personnel do all the work so that maximum leave and liberty can be granted, then the ship will simply not be ready to cope with casualties at sea.

The 3-M System is an integrated management system which provides for the use of available resources in performance of required maintenance and repairs. It sets forth procedures for collecting, processing, and analyzing data concerning the repair or maintenance and the material used. The objectives of the system are accomplished through two major subsystems: the Planned Maintenance System and the Maintenance Data Collection System.

PLANNED MAINTENANCE SYSTEM (PMS)

This system is designed to attain and maintain maximum operational efficiency of all shipboard equipment at all times, to reduce downtime of equipment, and to reduce maintenance costs. It is installed aboard all major U. S. Navy ships, and consists of a manual, a set of maintenance-requirement cards for each piece of equipment installed, and quarterly scheduled work cycles. The cards are so complete that they even list the appropriate instruction books, blueprints, tools, and hours required to perform the maintenance. Details of the system are available in NavPers 10788A as well as in the *PMS Manual.* These details are not reviewed in this Guide because they must be learned through experience. The program, whose procedures are so explicit that they can be followed by a user who has no management experience, is considered excellent. It contains only two pitfalls:

1. ship personnel sometimes falsely certify accomplishment of the maintenance procedures.
2. ship personnel are sometimes too inexperienced to perform the work required by the maintenance procedure and it is therefore not accomplished.

These pitfalls can be corrected by a proper management control system and an adequate training program, respectively.

In the authors' opinion, PMS is one of the most worthwhile advances in Navy management in the past two decades.

MAINTENANCE DATA COLLECTION SYSTEM (MDC)

The object of this system is to gather information on what resources (man-hours, material, and money) are spent on maintaining and repairing equipment. Thus, the system provides department heads with a means of planning, scheduling, and performing maintenance, because it details man-hour utilization data and thereby promotes effectiveness of personnel distribution and work assignment. The details of this system are subject to constant change, and it is therefore essential that shipboard personnel review OpNav-43 P2 (Revised) and other pertinent OpNav instructions relating to its use. Again, details of the actual workings of the system are not discussed in this Guide. Naval training commands offer short courses in both MDC and PMS. It is recommended that leading petty officers and shipboard officers attend these courses or, at least, read the system manuals and use the self-teaching "programmed instruction" courses. The major pitfall in MDC is simply that too much is expected of it.

The MDC program was, unfortunately, sold to the fleet on the basis of how individual ships would benefit, and many people expected the benefits to be tangible and immediate. In actuality, it is the authors' belief that the MDC system is an excellent device for substantiating resource requirements at the levels of the Secretary of the Navy and the Secretary of Defense. It also provides the Navy Supply System with a vital tool in determining what parts have high failure rates, thereby allowing it to direct the allocation of its dwindling funds to their replacement. Thus, although ships may still be frustrated by lack of operating funds and supply delays, progress is being made at high management levels and eventually the system will substantiate the true material needs of the Navy. This, of course, will happen only if each ship faithfully follows the procedures for submitting MDC reports.

It is difficult at times to sell shipboard personnel on the vague, optimistic returns of the MDC system, as opposed to their increased paper work, but it is a basic leadership responsibility for division leading petty officers, division officers, and the department head to insist that this job be carried out in an effective manner.

TYPES OF DATA OBTAINED BY THE 3-M SYSTEM

As a result of PMS and MDC, some important data has been accumulated and published. Much of it has been published and is obtainable through the Defense Documentation Center (DDC), an outstanding reference source which, unfortunately, too few people in government use properly. The following two DDC reports based on the results of ships reporting through the 3-M System, should be sufficient to convince personnel that the system does help the fleet.

1. AD 864259 "Steam Driven Turbo-Generators Maintenance Analysis Report," Sept. 1969.

SUMMARY RESULTS

Problem Areas for DLG-6 through DLG-15 Remarks

Turbine, EIC AP01—This component accounted for 25 percent of the maintenance costs. Its failure rate was the highest of all components.

Pump, EIC AP23—This component accounted for 19 percent of the maintenance costs.

Pump, EIC AP28—This component accounted for 14 percent of the maintenance costs.

Parts—Troubles with each component were associated with relatively few parts.

Maintenance Cost per Ship Configuration per Month were:

Labor (Computed at $10 per Hour)	$320.18
Parts	$144.43
Total Labor and Part Costs	$464.61

Equipment Reliability and Maintainability Indices

Failure Rate (Failures per 10^6 Hours of Operation)	2946

MTBF (Mean Time Between Failure) (Hours)	339
MTTR (Mean Time to Repair) (Hours)	15
Median Repair Time (Hours)	3.8
A_i (Inherent Availability)	0.9575
MTBPR (Mean Time Between Part Replacements) (Hours)	153
Percentage of Total Material Demands Deferred	32

Selected Component Failure Rates (Failures per 10^6 Hours of Component Operation)

Turbine, EIC AP01	1048
Pump, EIC AP23	449
Pump, EIC AP28	504

2. AD 839073 "Maintenance Comparison, 600 and 1200 psi Boilers in Destroyer Type Ships"

SUMMARY RESULTS

Total Maintenance Cost per 1000 Boiler Operating Hours

A. The 600 PSI boiler group has a lower total maintenance cost per 1000 boiler operating hours than the 1200 PSI boiler group.

B. The Foster Wheeler Boilers have a lower total maintenance cost per 1000 boiler operating hours than do the Babcock and Wilcox Boilers in the 600 PSI boiler group.

C. Within the 1200 PSI boiler group, the Combustion Engineering Boilers have the lowest total maintenance cost per 1000 boiler operating hours and the Babcock and Wilcox Boilers the second lowest.

D. Comparing all boilers, the Foster Wheeler 600 PSI Boilers have the lowest total maintenance cost per 1000 boiler operating hours and the Combustion Engineering 1200 PSI Boilers slightly higher.

Mean Time between Failure (MTBF)

A. The 1200 PSI boiler group has a higher MTBF than the 600 PSI boiler group.

B. The Babcock and Wilcox Boilers have the highest MTBF within the 600 PSI group.

C. The Combustion Engineering Boilers have the highest MTBF within the 1200 PSI group.

D. Comparing all boilers, the Combustion Engineering 1200 PSI Boilers have by far the highest MTBF and the Foster Wheeler 600 PSI Boilers the lowest MTBF. All other boilers have approximately the same MTBF.

There are many more categories for which such reports have been prepared. Appendix D contains extracts from reports for some of the major shipboard equipment: high-pressure air compressor, forced-draft blowers, the Mark 19 gyrocompass, and distilling plant. A ship armed with the knowledge contained in these reports should be well on her way to the goals of never missing a commitment and of winning the coveted engineering "E."

BASIC RESPONSIBILITIES FOR MAINTAINING AN EFFECTIVE 3-M PROGRAM

The *3-M Manual* allocates responsibility as follows:

> The *Division Officer* plays a key role in the shipboard 3-M organization. He is responsible to the department head for the effectiveness of the 3-M program within his division and assists the department head in scheduling maintenance required for the equipment under his cognizance. In fulfilling these responsibilities he is required to:
>
> 1. Ensure, by daily inspections, that weekly PMS schedules are in accordance with the departmental quarterly schedule and that the required maintenance is, in fact, being properly performed and reported.
> 2. Incorporate 3-M training into the divisional training plan and maintain up-to-date 3-M training records for his division.
> 3. Ensure that MDCS documents generated within his division are complete, accurate and promptly submitted.
> 4. Assist the 3-M Coordinator and the Departmental 3-M Assistant as necessary in all matters concerning 3-M within his division.

5. Conduct periodic meetings with all divisional work center supervisors and keep the department head informed of the status of 3-M within the division.

The Division Officer is required to attend a formal course of instruction in the 3-M System at one of the Fleet Training Centers. There he is taught the content, preparation and use of the various forms currently incorporated into the system. Thus equipped, he should be able to carry out his supervisory and training responsibilities.

The *Department Head,* assisted by an appointed Departmental 3-M Assistant and the Division Officers of his department, is responsible for the necessary coordination to bring about the effective operation of the 3-M System within his department.

The *Supply Officer* coordinates his department's efforts to best support the ship's maintenance effort.

3-M Coordinator. He is directly responsible to the Executive Officer for coordination and direct supervision of ALL FACETS of the ship's 3-M Program.

The *Executive Officer* is, in turn, responsible to the Commanding Officer for the overall organization and operation of the ship's 3-M program. In stating CNO policy concerning command relationships and responsibilities, the 3-M Manual is particularly emphatic—"It is emphasized that 3-M is not a permissive system and that the key to success is active command attention and aggressive supervision at all levels, from operational commander to the work center supervisor."

3-M INFORMATION RETRIEVAL

The *3-M Manual* describes the system's retrieval function as follows:

> The management and system information continually being fed into the system by fleet units using MDCS documents is retrievable in data printouts which are provided on a recurring or demand basis, as established by the "user" activity. Demand reports may be either of a summary or detailed nature, concerning any system, equipment or component which is

included in the 3-M System. They are produced to meet specific requests using only a specified portion of the accumulated 3-M data bank and can be requested by any level of management within the Navy Command structure. Recurring reports are produced on a periodic basis or requested by the "user", generally at or above the Fleet or Type Command level, who then distributes them to the subordinate level (s) of command desired. The most important of these reports at the shipboard level is the Current Ships Maintenance Project (CSMP). Depending on the accuracy of the submitted MDCS documents, the retrieved 3-M "product," the CSMP, can be the most effective management tool available to the ship or the most vexing piece of paper to ever arrive on board. A truly current CSMP will list every unaccomplished maintenance item aboard ship and indicate the repair level at which it is to be accomplished. It can greatly assist in the planning and coordination of the ship's maintenance effort. However, never has the old computer adage, "GI-GO" (Garbage In—Garbage Out), been more appropriate. Computers do not generate inaccurate data. If inaccuracies exist, their origin is quite obvious—the shipboard organization.

The Maintenance and Material Management (3-M) System is an integrated management system in which maintenance data are collected in individual work centers, reported through the Maintenance Data Collection Sub-System (MDCS), processed at computer installations ashore and afloat, and disseminated in the form of automatic data processing (ADP) machine printouts. A great deal of effort is expended in collecting and processing 3-M data. Constant supervision at the documentation level, i.e., the Division Officer, will ensure that the value of the returned "3-M product" justifies that effort.

CHAPTER 12

Shipyard Overhaul

A young ship's officer was heard to say, "This is the most traumatic experience of my life." Shipyard overhaul need not, however, have such an impact on the crew of a proud ship if they are properly prepared to embark upon this venture. It is a venture designed to restore and repair presently installed equipment, to outfit the ship with the most current weaponry, communications, and detection equipment, and to prepare her for extended operating periods at sea.

PREPARATION

In reality, preparation for shipyard overhaul commences as the ship departs from her latest overhaul period. Once out of overhaul, a work request file should be created and it should be maintained on a continuing basis. As many work requests as possible should be accomplished by ship's force during normal in-port upkeep periods, tender availabilities, and restricted availabilities. This will leave a set of "hard core" shipyard overhaul requirements and facilitate the type commander's screening of overhaul work requests.

It must be remembered that a shipyard's contract with the type commander and the various systems commands calls only for specific work to be accomplished. At no time, should ship personnel develop the attitude that everything on board will be overhauled or repaired. The only work that will be accomplished is that which is authorized and funded by cognizant authority. This authority *is not* vested in shipyard personnel at any level of the organization.

The various types of work done during an overhaul and the types of overhauls conducted are explained in *Navy Regulations,* 1948, Chapter 20. The following definitions have been excerpted from Section 3, of that chapter:

1. *Repair*—Work necessary to restore a ship or article to serviceable condition without change in design, materials, number, location, or relationship of the component parts.

2. *Alteration*—Any change in the hull, machinery, equipment, or fittings which involves a change in design, materials, number, location, or relationship of the component parts of an assembly regardless of whether it is undertaken separately from, or incidental to, or in conjunction with repairs.

3. *Alterations equivalent to repairs*—An alteration shall be considered equivalent to a repair when such alteration consists of:

(a) The use of different materials which have been approved for like or similar use and such materials are available from standard stock.

(b) The replacement of worn-out or damaged parts assemblies or equipments requiring renewal by those of later and more efficient design previously approved by the bureau concerned.

(c) The strengthening of parts which require repair or replacement in order to improve reliability of the parts and of the unit, provided no other change in design is involved.

(d) Minor modifications involving no significant changes in design or functioning of equipment but considered essential to prevent recurrence of unsatisfactory conditions.

Only the bureau exercising technical control over the article, or the authority to whom such technical control has been delegated by that bureau, shall designate an alteration as equivalent to a repair and approve it for accomplishment.

4. *Availability*—The period of time assigned a ship by competent authority for the uninterrupted accomplishment of work which requires services of a repair activity ashore or afloat.

5. *Restricted Availability*—An availability for the accomplishment of specific items of work by a repair activity, normally with the ship present, during which period the ship is rendered incapable of fully performing its assigned mission and tasks due to the nature of the repair work.

6. *Technical Availability*—An availability for the accomplishment of specific items of work by a repair activity, normally with the ship not present, during which period the ship's ability of fully performing its assigned mission and tasks is not affected by the nature of the repair work.

7. *Regular Overhaul*—An availability for the accomplishment of general repairs and alterations at a naval shipyard or other shore based repair activity, normally scheduled in advance and in accordance with an established cycle.

8. *Interim Overhaul*—A scheduled availability of not more than one-half the duration of a regular overhaul, for the accomplishment at a naval shipyard or other shore based repair activity of necessary repairs and urgent alterations. Normally, an interim overhaul will be scheduled approximately midway between regular overhauls.

9. *Supply Availability*—A period of time assigned a ship by competent authority for the uninterrupted accomplishment of a supply overhaul. A supply availability is normally scheduled to coincide with a regular overhaul.

10. *Supply Overhaul*—The work involved in the

purification and adjustment of on-board stocks and records to bring them into conformance with prescribed allowances or other stockage objective criteria.

11. *Voyage repairs*—Emergency work necessary to enable a ship to continue on its mission and which can be accomplished without requiring a change in the ship's operating schedule or the general steaming notice in effect.

12. *Upkeep period*—A period of time assigned a ship, while moored or anchored, by competent authority for the uninterrupted accomplishment of work by the ship's force or other forces afloat.

Although engineering officers aboard ship may not be aware of it, the shipyard actually begins preparing for her overhaul, from a planning and design standpoint, when it receives the systems commands' advance alteration letters. This is some three or four months prior to the scheduled overhaul date. It is, therefore, essential that the proposed alterations be reviewed in detail by the ship and that those which are completed, partially completed, or cannot be completed be listed in a letter for submission to the cognizant systems commands. Remember that the lead time is essential in order to allow the shipyard to draw up proper plans for issuance to its repair department. If the shipyard has the time and the money, it will send an advance "ship check" team to the ship. The team must be given the fullest cooperation, but it must be remembered that nothing that is said in the course of such inspection will, or should, be considered by the shipyard as work authorization. Since funds are normally austere, cognizant authority may not authorize all the repairs or alterations in the exact manner they are desired by the ship's officers or by the shipyard. This may be unavoidable, and the crew must be psychologically prepared for the deletion of some of the proposed work. At times, the ship's crew will be required to share in the accomplishment of certain work, and, in some cases, to do the entire job themselves. Economics are, and must be accepted as, a prime consideration in the amount of work that will be accomplished during overhaul. Leading members of the crew should be given some instruction in funding procedures, in order to prevent serious misunderstandings between them and shipyard personnel. This knowledge also strengthens the ship's bargaining position at the arrival conference.

The key to getting ship repairs done, lies in the preparation of the work request, which should be written concisely, but should be explicit and complete. It should contain accurate data as to the type failures, the number of repetitive failures, the applicable blueprint numbers, the known replacement part numbers, and the names of ship personnel knowledgeable concerning the requested repair. Proper preparation of the work request can be one of the ship's most valuable assets at the arrival conference.

ARRIVAL CONFERENCE

Once the ship's work requests, work priority list, and systems commands' alteration letters have been forwarded by the ship to the proper authorities, with a copy to the shipyard in each case, the shipyard will then evaluate the repairs and submit cost estimates to the type commander and cognizant systems commands. At this point, the shipyard, type commander, representatives of cognizant systems commands, and ship's force personnel will hold an arrival conference. The shipyard will present, at this time, a detailed work booklet stating the cost of each work item, and the type commander will assign responsibility for its accomplishment. Often, the type commander will assign a large work load to be done by the ship's force. The ship's officers must remember that the crew will not only be busy doing previously scheduled repair work, but many of them will be involved in fire and security watches as well as retraining in fleet schools. In addition, attrition due to transfer and retirement will lessen the work force available during the period of the overhaul. A "can do" attitude is fine, but a realistic approach must be adopted in accepting ship work. Be prepared to discuss with the type commander the ship manpower requirements in relation to the work he has assigned. If the ship has planned properly, his representative will take these plans into consideration. This will lead to a realistic assignment of work and benefit the ship and the shipyard. Representatives of cognizant systems commands will also authorize work at this time. When the arrival conference has been completed, the ship will be scheduled for offloading ammunition, fuel, stores, and equipage in preparation for entering the shipyard.

SHIP ARRIVAL

After the work has been authorized by cognizant authority, the shipyard will prepare detailed job orders and issue them to the ship on her arrival in the yard. It is essential that responsible ship personnel review each job order to ensure that the work contained therein complies with the ship's work requests, as agreed upon by the funding authority at the arrival conference.

Shortly after arrival in the shipyard, ship personnel will be invited to attend a shipyard briefing. Senior enlisted grades are urged to attend along with all officers. During the briefing, information will be presented relating to

1. the organization of the shipyard, with particular emphasis on its planning and production departments
2. the processing of requisitions for material
3. general relations between ship personnel and the yard's supply department
4. the issuing of portable tools
5. the handling of "assist ship force" work requests
6. the procedures for processing of ship-to-shop work
7. the number of fire and security watches that will be required.

At the same time, the ship will be provided with the names of those shipyard personnel equipped to handle questions emanating from the ship, and a list of telephone numbers to be used in case of fire, flooding, oil spill, or the need for gas-testing, police, or ambulance service.

A ship superintendent will be assigned to the ship and will serve as official liaison between the ship and the shipyard workers. He should be used by the ship when it is necessary to obtain advice on shipyard procedures, express displeasure at shipyard work, and settle scheduling conflicts. Remember that he is the direct representative of the shipyard production officer for the ship and, as such, has full authority over all shipyard work on the ship to which he is assigned.

It is essential that ship's force establish good relations with shipyard personnel. This is not achieved by conducting a give-away program. It is accomplished by establishing a mutual respect for one's fellow man. A successful overhaul depends on the ship's company working together with the shipyard personnel to obtain a mutual understanding of their respective problems and the solutions thereto.

SHIPYARD WORK ROUTINE

GENERAL WORK

Once work on the ship has commenced in earnest, there are certain key procedures which, if followed, will help to produce a successful overhaul:

1. Establish ship's force work schedules similar to those prepared by the shipyard for shipyard work.
2. Review shipyard work schedules. Have weekly progress meetings with the ship superintendent to ensure that shipyard work is on schedule. This helps expedite shipyard work and also points up possible areas of conflict between ship and shipyard schedules. Remember that the reason for doing everything possible to keep the work on schedule is to permit early checkout of systems for which both the ship and the shipyard personnel have responsibilities.

3. Strip the ship of all equipage that can be stored ashore. This will forestall the possibility of damage. The shipyard will assist in locating proper storage facilities.
4. Supervise "open and inspect" work. If such inspection discloses the need for corrective actions, be sure that supplemental work requests are forwarded to the funding authority as soon as the inspection has been completed.
5. Review all work items. Submit additional supplemental work requests when work not originally scheduled is required. Supplemental work should be submitted as early in the overhaul period as practical. More effective results will accrue to the ship if the shipyard does not have to use overtime to get work accomplished.
6. Preventive maintenance programs should be maintained on items not undergoing repairs. As an example, electric motors should be turned over at least once a week.
7. Ensure that the ship is airtight during sandblasting operations. This will not only prevent dirt accumulation but will prevent serious damage to open systems and pieces of equipment.
8. Keep a check on the progress of ship-to-shop work. Have men available to witness tests and to accept completed equipment.
9. Insist on a clean ship. This can be accomplished through a mutual effort by ship's force and shipyard personnel. It should be coordinated through the ship superintendent. *Keep pushing for a clean ship.* The result will be less

damage to newly installed or open equipment, less likelihood of fire, and the elimination of numerous hazards to personnel.

10. Inspect ship systems and follow shipyard work carefully. Understand your men and determine who you can rely upon. Investigate all derogatory reports on shipyard work to ensure that they are complete and accurate prior to alerting your commanding officer. Try to have installed systems placed in operating condition promptly, and certainly prior to scheduled engineering trials.

The above list should assist in getting the ship off on the right course. To list all of the intricacies of shipyard overhaul could well be the subject of an entire textbook, and knowledge of many of them can be gained only from experience. However, adherence to the above guidelines, coupled with those to follow, will bring the ship to the culmination of repairs and the preparation for trials and completion of overhaul.

DRY-DOCK WORK

Details for ships to be dry-docked are normally provided to the ship by the shipyard docking officer prior to docking. The following are some of the items considered to be of specific importance:

1. It is normally required that all fuel, gasoline, ammunition, and other incendiary devices be removed from the ship before she enters the dry dock.
2. The ship should be brought to a condition of zero list and minimum trim.

3. The docking plan should be examined and liquids distributed in accordance with the prescribed plan.
4. The shipyard inspectors should be accompanied on their underwater hull inspection. Ascertain whether scheduled repairs are those desired by the ship. If this is not clear, submit supplemental work requests to ensure that the desired work is covered.
5. While in dry dock, the following precautions and regulations should be observed:
   a. After landing on the blocks, do not make any changes in weight or distribution of liquids without the consent of the docking officer.
   b. Keep all wing tank cross connections closed, in order to preserve maximum stability during docking and undocking.
   c. Do not rotate propellers, unless requested to do so by shipyard personnel, and then only with the consent of the docking officer.
   d. Do not discharge oil, garbage, wastes, or trash into the dry dock.
   e. See that all personnel entering the bottom of the dry dock wear safety helmets and goggles.
   f. Do not allow ship's force to work over the side.
   g. Remember that the ship is responsible for policing the dock, altars, and caisson.
   h. Keep all loose equipment at least two feet away from the side of the ship.
   i. Do not permit any water washdown. Remember that there is a lot of electrical equipment on the dock floor.

j. Maintain a log of all weights added or removed and submit it to the docking officer prior to undocking.
k. Be sure that all ship's work is completed and that it, along with shipyard underwater work, is certified prior to undocking.
l. When the dry dock is being flooded, station ship's personnel at all underwater openings to report immediately if any leaks develop. It is the responsibility of the commanding officer to see that this precaution is taken.

When the dry-dock work is completed and all ship systems have been checked out satisfactorily, the final phase of shipyard overhaul consists of trials.

SHIPYARD TRIALS

When overhaul of the ship has progressed steadily toward completion in all areas, if strict adherence to schedule has been maintained, it is time for trials. Do not, however, be forced into conducting trials prematurely. At this stage of overhaul, most mishaps are the result of human, rather than mechanical, failure. Ensure that the crew is rested and trained in their duties by use of familiarization on-station drills. If you delay too long, you will delay completion, and the shipyard will be required to go to a costly overtime basis to complete the ship on schedule.

By following a few basic guides and procedures, the engineer officer can enhance the ship's ability to conduct timely and successful engineering trials. Success in this area, in turn, contributes to timely and economical completion of the ship's overhaul and, perhaps, to the training of future operating engineers. Much of the following information applies to all steam-driven ships.

LIGHT-OFF TRIAL

Initial Ship Check

The first major engineering trial during overhaul is light-off. This normally consists of firing boilers and testing all auxiliary machinery, including ship service generators. In many cases, main plant machinery is tested but main engines always remain secured. Since much of the machinery has either been inactive for at least two months or has undergone extensive shipyard overhaul, this is a critical phase. Three to five days before light-off, the ship's engineer officer and the ship superintendent should jointly inspect the engineering plant to make sure that:

1. Feedwater has been ordered in ample supply to fill tanks and boilers for initial light-off.
2. All piping systems have either been completed or that open lines have two-valve protection or are blanked, and that all relief-valve-escape piping has been installed.
3. Fuel oil has been ordered and is in service tanks ready for service-pump suction, and the fuel-oil-transfer system and fuel-oil heaters have been checked.
4. Lubricating oil has been ordered and installed in all machinery.
5. Cooling water is available to the lubricating oil coolers.
6. All valves in the systems are closed. (These will

be opened progressively during the light-off procedure.)

7. Main and auxiliary circulators have been tested and condensers have been checked for external leakage. (All condensers must be vented before starting pumps.)
8. Forced-draft blowers have been checked and are prepared to run. Normally, one blower is run on dock steam until sufficient boiler-steam pressure has been raised.
9. Ship's service generators have been prepared to run, and their oil level, coolers, thermometers, hand pump, etc. have been checked.
10. Air ejectors have been checked with respect to nozzle installation and gasket tightness.
11. All gages and gage lines have been installed.
12. All thermometers have been installed.
13. All electric motors have been checked for grounds, have been turned by hand, and then test-run on shore power.
14. Evaporators are ready to provide feedwater for future steaming requirements. Commercial feedwater is quite expensive.

Light-Off

After the initial ship check has been made and all items are satisfactory, the light-off can take place. During light-off, the procedure is approximately as follows:

1. Have boilers run down to operating level.
2. Check for good feedwater.
3. Remove stack covers.

4. Have all valves closed before starting. Open valves slowly to avoid blowing out flange gaskets or rupturing lines.
5. Use in-port size sprayer plates for light-off and rotate burners to get an even heating of furnace refractories.
6. Exhaust all steam to the atmosphere until the auxiliary plant is ready to receive it.
7. When steam pressure on the boiler exceeds dock pressure, secure shore steam and bring in auxiliary steam from the boiler. Direct steam to the machinery plant.
8. When the boiler has a full head of steam, set safety valves. Line up the auxiliary feed pump to a cold feed bottom, ready to supply feedwater to the steaming boiler.
9. Light off an auxiliary plant. Shift the exhaust steam flow from the atmosphere to the deaerating feed tank and auxiliary condenser. Check out auxiliary air ejectors, deaerating feed tank, condensate system, and feed system.
10. Make sure that valves are opened only at the direction of the man in charge of the light-off operation. Each section of pressurized line must be checked for leaks, and leaks should be corrected promptly.
11. When the auxiliary lines, desuperheater line, auxiliary exhaust line, and the condensate and feed systems have been checked, test the main steam line up to the guarding valve.
12. When all systems and auxiliary machinery have been checked and determined satisfactory, start the ship's service generators and commence set-

ting overspeed trips. Conduct heat and vibration runs on salt-box loading. Once generators have been checked satisfactorily by salt-box loading, they should each take the ship's load separately and then in parallel.

13. Light off evaporators and commence making feedwater.

During this first light-off period, every piece of machinery and all piping should be thoroughly tested. Machinery should continue to be run as often as possible before sea trials to prove reliability and to permit repair time, if failure occurs.

DOCK TRIAL

The second major engineering trial is the dock trial, which consists of actually turning main engines up to shaft speeds of approximately five knots while the ship is dockside. Speed is brought up gradually with one engine going ahead and the other astern, and then shaft rotation is gradually reversed. Before the dock trial, checks should be made to ensure that

1. the area surrounding the propellers is clear.
2. some freedom to permit ship surge is available at bow and stern.
3. mooring lines and wires are rigged properly.
4. camels are secured to dock pilings.
5. stern-tube packing glands have been slacked and proper cooling water has been supplied to the gland.
6. spring bearings have been lubricated.
7. the brow has been removed.

During the actual dock trial, all machinery and piping systems should be tested and operation proved to the satisfaction of both the ship and shipyard personnel.

FULL-POWER MACHINERY SEA TRIAL

The final major engineering trial is the full-power machinery sea trial, which is conducted at or near full power after a normal warm-up and test period at sea. Naturally, a prerequisite is that all main propulsion machinery and associated auxiliaries have been tested satisfactorily at the dock trial. Since experience has proven that failure to achieve full power is usually caused by improper operation, specific machinery problems must be clearly understood, and machinery test status reports must be recorded.

Specific Machinery Problems

Proper operating temperatures and pressures are often not obtained during full-power, dock, and/or light-off trials. Typical malfunctions encountered during full-power machinery sea trials are:

BOILER (LOW STEAM-DRUM PRESSURE)

Fuel-oil temperature and pressure to burners is insufficient.
Wrong size or dirty burner tips are being used.
Atomizer setting is improper.
Water level in steam drum is too low.
Steam demand is excessive.
Instrumentation is faulty.

CONDENSER (LOW VACUUM)

Gland-seal steam is not at design pressure.

Insufficient cooling water is being supplied to the condenser.

Condensate pumps are not running properly.

Air ejectors are malfunctioning.

Air is leaking into the condenser, exhaust trunks, etc.

Condenser is flooded or dirty.

Injection temperature is high.

Water in feed tank on which vacuum drag is being taken is inadequate.

Loop-seal filling valve is open.

Bypass valve on drain tank is open.

DEAERATING FEED TANK

Water level has dropped, because a condensate pump is not operating properly or a recirculating valve is open.

Water level has increased because the feed booster pump is not operating properly or a make-up feed valve is open.

AIR EJECTOR

Nozzle capacity is so great that one nozzle for each stage will serve at full-power operation. The design steam pressure required by the nozzle must be maintained. If the pressure is carried below normal, the ejector operation will become unstable and destroy condenser vacuum. If carried much above design, a reduction in vacuum can be expected because the air ejector and intercondensers will be overloaded.

Other possible causes of faulty operation are dirty steam strainers, improper adjustment of the unloading valve, or insufficient cooling water to the inner and after condenser.

LOW FEEDWATER PRESSURE

A feed-check valve on an idle boiler is open.
Feed-pump regulator is not working.
Feed pump has vapor in it or is air-bound.
Feed booster pump pressure is below design.
Pump clearance is faulty or a casing has eroded.
Valve in recirculating line is closed.
Feed pump is operating below rated rpm.

LOW AIR PRESSURE TO BOILERS

Forced-draft blower is not operating properly.
Air registers are faulty.

IMPROPER LUBRICATING OIL TEMPERATURE

Bearing inlet oil temperature should be 120°-130°F. from the cooler. Temperature is controlled by increasing or decreasing the circulating water through the lube-oil cooler. Operating the system with oil inlet to the bearings below normal may increase the temperature rise in some bearings and result in their showing higher outlet temperatures than they would if supplied with oil at the correct temperature.

IMPROPER FUEL-OIL TEMPERATURE

Temperature at the fuel-oil service pump discharge is not being maintained at design pressure. Pressure at burner manifold should be about 300 psi. Temperature should be as set forth in NavShips *Technical Manual,* Chapter 9550.

LUBRICATING-OIL PRESSURE

A sudden increase at the pump usually indicates a clogged strainer.

A loss of pressure is normally caused by pump or line failure.

CENTRIFUGAL PUMP OPERATING DIFFICULTIES

Air leaks in suction or stuffing boxes.

Pump not up to speed.

Impeller partially clogged.

Suction-water low.

Mechanical defects, wearing rings worn, casing gaskets defective, etc.

Low output pressure.

Pump is not up to speed.

Air leaks. (In general, this applies to reciprocating and rotary pumps.)

Incorrect discharge valves in manifold are open.

Loss of capacity or decreased flow rate after starting.

Air leakage in suction line.

Stuffing-box water seal is plugged.

Strainer is fouled.

Air is in water.

HOT BEARINGS

Lubrication is improper or insufficient.

Grit or dirt in the lubricant.

Bearings are out of line.

Bearings are improperly fitted.

Oil lines are obstructed.

Normal bearing inlet temperature is from 120° to 130°F., and normal outlet temperature is 140° to 160°F.

Test Status Reports

Machinery test status is difficult to maintain by memory during major engineering trials. Therefore, the use of forms such as those shown in Figures 9 and 10 is suggested. They provide engineering personnel with an exact and timely status of machinery tests from commencement of light-off until completion of full-power machinery sea trial. They also list typical machinery items that must be checked out during both dock trial and full-power machinery sea trial. A date in the appropriate box indicates that the machinery item was tested and proved satisfactory during the light-off, dock-trial, and sea-trial phases. Space is left so that additional items can be inserted; machinery item numbers are blank to make the form more versatile. A column is provided to indicate whether repair responsibility belongs to the ship's force or to the shipyard.

COMPLETION OF TRIALS

If, upon completion of any of the above trials, any malfunction has come to light, the engineer officer should prepare a set of detailed discrepancy cards defining the corrective action required. Once discrepancies have been corrected successfully, there should be little, if any, work left aboard ship under the cognizance of the shipyard. Alert the ship superintendent to any work that is still outstanding. The shipyard will want to finish the ship and get on to the next scheduled overhaul.

USS ____________

MACHINERY CHECK LIST

Date tested satisfactorily

Light Off	Dock Trial	Mach. Sea Trial	Engineroom	Repair Responsibility
			#____ High Pressure Turbine	
			#____ Low Pressure Turbine & Astern Turbine	
			#____ Cruising Turbine	
			#____ Main Condenser (____ Vacuum)	
			#____ Main Air Ejector	
			#____ Deaerating Feed Tank	
			#____ Main Circulating Pump	
			#____ Ship's Service Generator	
			#____ Auxiliary Air Ejector	
			#____ Lubricating Oil Purifier	
			#____ Lubricating Oil Cooler	
			#____ Lubricating Oil Service Pump	
			#____ Lubricating Oil Service Pump	
			#____ Main Feed Pump	
			#____ Main Feed Pump	
			#____ Main Feed Booster Pump	
			#____ Main Feed Booster Pump	
			#____ Main Condensate Pump	
			#____ Main Condensate Pump	
			#____ Fire & Bilge Pump	
			#____ Fire & Flushing Pump	
			#____ Auxiliary Condenser	
			#____ Auxiliary Condensate Pump	
			#____ Auxiliary Feed Booster Pump	
			#____ Auxiliary Circulating Pump	
			#____ Distilling Plant (,000) gal.	
			#____ Low Pressure Air Compressor	
			Diesel Oil Purifier	
			Diesel Oil Supply and Transfer Pump	
			#____ Fresh Water Pump	
			#____ Fresh Water Pump	
			#____ Shaft Driven Lubricating Oil Service Pump	
			Additional Items	

Date ____________

Figure 9—Sample Engineroom Machinery Check List

USS ______________

MACHINERY CHECK LIST

Date tested satisfactorily

Light Off	Dock Trial	Mach. Sea Trial	______________ Fire Room	Repair Responsibility
			Final Hydrostatic Test _____ Boiler	
			Satisfactory Safety Valve Setting A___ B___ C___ D___	
			Final Hydrostatic Test _____ Boiler	
			Satisfactory Safety Valve Setting A___ B___ C___ D___	
			# _____ Forced-Draft Blower	
			# _____ Forced-Draft Blower	
			# _____ Forced-Draft Blower	
			# _____ Forced-Draft Blower	
			# _____ Fire & Bilge Pump	
			# _____ Emergency Feed Pump	
			# _____ Fuel Oil Transfer and Booster Pump	
			# _____ Port and Cruising Fuel Oil Service Pump	
			# _____ Fuel Oil Service Pump	
			# _____ Fuel Oil Service Pump	
			# _____ Fuel Oil Hand Pump	
			# _____ HP Air Compressor	
			# _____ Fuel Oil Heater	
			Additional Items	

Figure 10—Sample Fireroom Machinery Check List

READY FOR SEA (RFS)

Upon completion of the overhaul, the ship will have to reload stores, ammunition, fuel, etc., and plans for doing so should be made well in advance.

Do not wait until the last day of the overhaul to make arrangements. The key to success is *foresightedness.*

When the ship is finally back at sea, do not forget the shipyard. Keep a check on the items that have been repaired. Keep operating records up to date. The commanding officer should report to the shipyard, in the form of constructive criticism, any faulty workmanship, poor installation procedures, and operating difficulties. The shipyard will be happy to make good any poor work as soon as the ship's operating schedule permits.

SUMMARY

If you now have the feeling that shipyard overhaul is a complicated period involving a great deal of work, then this chapter has served its purpose. The success or failure of shipyard overhaul depends to a great extent on mutual cooperation between the ship's force and shipyard personnel; enthusiastic support will always work toward this end. The data contained in this chapter is in no way a complete analysis of the overhaul period. It is designed to give a general overview of the problems that will be encountered. Details can be found in fleet, force, and type commanders' regulations, NavShips *Technical Manual,* and shipyard regulations. Most of these publications are available aboard ship; those that are not, will be provided to you prior to the start of overhaul preparation.

Remember—get the most from the overhaul. *There are no shipyards at sea.*

CHAPTER 13

Damage Control

Damage control encompasses the administration and training of relatively large numbers of men, plus the operation and maintenance of a complex array of equipment and machinery. This is a challenging task made even more difficult by an attitude of complacency which tends to set in after long accident-free periods. In this respect damage control is much like life insurance because its importance is never fully appreciated until disaster strikes. Establishing the proper attitude, especially in peacetime, is perhaps the most difficult and urgent task associated with damage control.

It is almost three decades since a major sea battle involving extensive ship damage has occurred. This has in some measure contributed to a dangerous lack of appreciation for the modern-day applicability of damage control. One typical school of thought, which is a product of the nuclear age, is the all-or-nothing concept. The idea that a ship undergoing a nuclear attack will either be instantly reduced to a charred cinder or come out completely unscathed is an oversimplified philosophy, at best.

The force that motivates a crew to strive for excellence in damage control is supplied automatically in time of war. Man's basic instinct for survival creates an instant desire in every sailor to learn the rudiments of how best to keep his ship afloat—he needs no further motivation.

In peacetime the learning and practice of damage control is largely reduced to a kind of "gamesmanship." To whip up enthusiasm and provide incentive by competition, both within and without the ship, is encouraged. Shipboard damage-control leaders are hard pressed to convince each and every member of the crew that damage control is an important and vital part of their training. It is a constant battle. To win that battle there are a number of practical problems which must be identified and solved. Among the first is an understanding of the objectives of damage control.

OBJECTIVES OF DAMAGE CONTROL

1. The three basic objectives of shipboard damage control are:

a. To take all practicable preliminary action, before damage occurs, such as maintenance of watertight and fumetight integrity, preservation of reserve buoyancy and stability, removal of fire hazards, and upkeep and distribution of emergency equipment.

b. To minimize and localize such damage as does occur, by such actions as control of flooding, preservation of stability and buoyancy, combating fire, and first-aid treatment of personnel.

c. To accomplish, as quickly as possible, emergency repairs or restorations, after the occurrence of damage, by such actions as supplying of casualty power, regaining of a safe margin of stability and buoyancy, reinforcement of damaged structures, and manning of essential equipment.

2. Upon its effectiveness may depend the ship's ability to inflict punishment upon and destroy the enemy or to perform any other assigned mission. It is essential that every member of the ship's company recognize his responsibility and its importance. Damage control must be considered an "offensive", as well as defensive function. [NavShips *Technical Manual,* Chapter 9880.]

SCOPE OF DAMAGE CONTROL

Damage control is concerned not only with battle damage but also with nonbattle damage such as fire, collision, grounding, or explosion. It may be necessary in port as well as at sea, and may involve the use of personnel and facilities of an undamaged ship. [NavShips *Technical Manual,* Chapter 9880.]

Too often damage control is considered the exclusive province of the engineers aboard ship. This is a natural assumption since they are the people most directly involved with the maintenance and repair of equipment, as well as the establishment of standards of training. Furthermore, the Damage Control Assistant (DCA) is usually a member of the Engineering Department, thereby lending additional credence to the myth that damage control is strictly an engineering function. It must be emphasized from the very beginning that damage control is an all-hands evolution, and the DCA functions to organize and coordinate this all-hands effort.

DUTIES OF THE DAMAGE CONTROL ASSISTANT

GENERAL DUTIES

The damage control assistant of a ship, when assigned, shall be responsible, under the engineer officer, for establishing and maintaining an effective damage control organization, and for supervising repairs to the hull and machinery, except as specifically assigned to another department or division. [*Navy Regulations,* 1948, Article 0955.]

SPECIFIC DUTIES

The damage control assistant, under the engineer officer, shall be responsible for:

1. The prevention and control of damage, including control of stability, list, and trim. He shall supervise placing the ship in the condition of closure ordered by the commanding officer. He shall insure that appropriate closure classifications are assigned and conspicuously marked upon or adjacent to the objects to which they apply. He shall coordinate and supervise the carrying out of prescribed tests of compartments and spaces for tightness. He shall prepare and maintain bills for the control of damage and stability, and shall insure that correct compartment check-off lists are posted.
2. The training of ship's personnel in damage control including fire fighting, emergency repairs, and nonmedical defensive measures against gas and similar weapons.
3. The operation, care, and maintenance of auxiliary machinery piping, and drainage systems not assigned to other departments or divisions, and of the shop repair facilities; and the repair of the hull and boats. [*Navy Regulations,* 1948, Article 0956.]

To do his job properly the DCA must constantly cross departmental lines. His domain is the entire ship. Watertight integrity applies to the Operations and Weapons departments as much as it applies to the Engineering Department. The DCA plays a unique role in that he must take personnel from many other departments and teach them to function effectively as members of a repair party, of a fire-fighting team, or as Division Damage Control Petty Officers (DDCPO).

THE DIVISION DAMAGE CONTROL PETTY OFFICER

The Division Damage Control Petty Officer (DDCPO) is one of the key men in the shipboard damage-control organization. A ship without a full supply of good division damage-control petty officers will have a difficult time getting its damage-control program to progress beyond the mediocre stage. The DDCPO is the one man who keeps the whole division, and consequently the whole ship, actively engaged in the day-to-day business of damage control.

The following is extracted from the Engineering Department *Organization Manual* of a DLG-26 class ship:

> 1. The Division Damage Control Petty Officer, under the Division Officer, is responsible for the routine maintenance of the damage control equipment contained in divisional spaces and for the proper setting of the material condition ordered by the Officer of the Deck. He keeps the Damage Control Assistant informed of the status of damage control equipment and he assists the Division Officer in the preparation of required reports.
>
> 2. Specifically he shall:
>
> a. Acquaint himself with all phases of the ship's damage control organization and damage control procedures.
>
> b. Insure that the Damage Control Compartment Check-Off Lists for all spaces under his cognizance are properly filled out in accordance with the Master Compartment Check-Off Lists.
>
> c. Assist in the instruction of division personnel in damage control, fire fighting and NBC defense procedures.
>
> d. Supervise the setting of specified damage control material conditions within divisional spaces and make required reports (unless this duty is specified elsewhere).

e. Insure that all doors, hatches, scuttles, air ports, vent cover operating mechanisms, dogs, knife edges and rubber gaskets are in good operating condition within their assigned spaces.

f. Weigh portable CO_2 bottles, inspect and test damage control and fire fighting equipment and prepare required reports for approval by the division officer in accordance with current ship's instructions.

g. Insure that all equipment at fire stations is properly hooked up and that all damage control fixtures are secured and operable.

h. Insure that all fog head orifices are free from bright work polish, dirt and salt water corrosion.

i. Insure that all battle lanterns, dog wrenches, spanners and other damage control equipment are in place and in usable condition in all divisional spaces.

j. Insure that all compartments, piping and damage control and fire fighting equipment are properly stenciled or identified.

k. Insure that all damage control label plates are legible and free from paint.

l. Insure that all sound powered phones are properly made up, jack box covers are in place and storage boxes are not filled with foreign objects.

m. Insure that all ventilation systems throughout their assigned areas are cleaned in accordance with current ship's instructions.

n. Insure the posting of safety precautions, operating instructions and other pertinent instructions in required divisional spaces, in accordance with U. S. Navy Safety Precautions (OPNAV 34P1) and Type Regulations.

o. Assist the division officer in inspection of divisional spaces for cleanliness and preservation and assist in preparing the required reports.

p. Insure that all electrical kickpipes and stuffing tubes are plugged if not in use.

q. Conduct daily inspection of divisional spaces for the elimination of fire hazards and flooding conditions (unless this duty is specified elsewhere).

r. Perform such other duties with reference to damage control and maintenance of divisional spaces as may be directed by the Division Leading Petty Officer and/or Division Officer. All other requirements should be channeled through the chain of command.

3. The Division Damage Control Petty Officer reports to:

a. The Division Officer for the performance of all specific duties and responsibilities.

b. The Damage Control Assistant to inform him of the status of damage control equipment.

All of this looks very good on paper, but what does it mean? It means a tremendous amount of work that cannot possibly be accomplished properly by a single individual. To prosecute the above duties in an efficient manner takes a conscientious petty officer who is attentive to detail and thoroughly indoctrinated in the fundamentals of good preventive damage control. This does not happen automatically. He must be selected, trained, and supported in his efforts, and he must be convinced that he is making a worthwhile contribution to the operational effectiveness of his ship. The choosing of good petty officers for the position is extremely important: some officers may have a tendency to use the best enlisted men in positions where their day-by-day performance is more obvious, and to designate damage-control responsibilities to less efficient petty officers.

Just as in casualty control, there are two distinct types of damage control: preventive and corrective.

PREVENTIVE DAMAGE CONTROL

Preventive damage control is associated with those measures that can be taken *before* damage occurs, and it is the foundation upon which successful corrective damage control is built. Most of the duties of the DDCPO are centered around the maintenance of an effective preventive damage-control program.

On first inspection, preventive damage control may appear to be an overwhelming task. There are so many compartments, hatches, fittings, and systems packed into a modern warship—how does one keep track of everything? It isn't easy, but neither is it impossible. The 3-M System has simplified matters somewhat by incorporating most of the care and maintenance requirements of damage-control equipment into the system. This precludes the need to produce voluminous quantities of check-off lists, instructions, and reports, as was the practice in the past.

One of the most vital prerequisites of a good preventive damage-control program is education. Frequently it is assumed that a commissioned officer or senior petty officer just naturally knows a great many things about damage control. This is an erroneous assumption. Almost all officers and enlisted men, with the exception of hull maintenance technicians, have much to learn about the "nuts and bolts" of preventive damage control. Some questions that might be asked are:

1. What are spanner wrenches and where are they normally stowed?
2. How often is a portable CO_2 bottle supposed to be weighed and why?
3. How can a compartment check-off list be checked for accuracy, and what are the procedures for making a correction to the list?
4. What is the proper method for stowing casualty power cable, and why is it stowed that way?
5. What is the proper method for stowing fire hose, and why is it stowed that way?
6. How are compartments and passageways numbered aboard ship?

These are just a few seemingly simple questions, and most Navy men do know the answers in general terms. But general terms are not sufficiently accurate when it comes to taking corrective action. Each officer and enlisted man needs to have detailed knowledge in the area of preventive damage control. For example, it is counter-productive for a division officer conducting an inspection of his spaces to look at a fire hose and conclude in the presence of others that it is probably stowed properly. It is either stowed properly or it isn't, and he should know the difference. Furthermore, he should insist that his subordinates know the difference as well, and that they take corrective action if needed. This kind of attention to detail, supplemented with a positive follow-up procedure, is the most effective way to sustain a successful preventive damage-control program.

CORRECTIVE DAMAGE CONTROL

There are very few authorities on the subject of corrective damage control remaining in the Navy today. As was pointed out at the beginning of this chapter, it has been over 25 years since the last major sea battle involving extensive ship damage. In spite of this lack of experienced personnel, the Navy has a vast store of information on the subject of corrective damage control which has been gathered from ships that survived severe battle damage, major accidents or natural disasters at sea. Much of this knowledge has been incorporated into NavShips *Technical Manual* and into the curricula of refresher training and damage-control schools. Looking at it from a practical point of view, corrective damage control is not the sort of thing in which the average sailor can ever expect to gain a great deal of experience, unless he happens to be a very unlucky fellow. But the mistakes and successes of the past have provided some valuable lessons to rely upon, should the need arise. Such a fundamental principle as the absolute requirement to immediately establish fire and flooding boundaries in damaged areas of the ship is the product of some disastrous experiences. Progressive flooding or an uncontained fire will ultimately spell doom for any ship. An intimate knowledge of a ship's compartmentation and layout has proven time and again to be one of the most vital factors affecting human survival in a heavily damaged vessel. The ability to establish communications at or near the scene of damage is another basic necessity.

No attempt will be made here to go into the details of corrective damage control. That subject could fill an entire book—and has. NavShips *Technical Manual,* Chapter 9880, Sections I and II, provides some excellent guidance on Stability and Buoyancy and Practical Damage Control. The *Manual,* combined with an individual ship's Damage Control Book and Diagrams, plus the numerous training courses, books, and pamphlets published by BuPers and other Systems Commands provide a wealth of information on every aspect of damage control. A new system entitled Personnel Qualification Standards (PQS) is now being implemented in the fleet and one of its primary goals is to provide *all hands* with the fundamental knowledge and skills of shipboard damage control.

To illustrate the fact that damage to ships occurs in peacetime as well as in war, the following information has been extracted from the *Ships Safety Bulletin,* a monthly bulletin prepared by the U. S. Naval Safety Center at Norfolk, Virginia:

Damage	*Estimated Repair Costs*
Aircraft Carrier—FIRE Air Group compartment completely burned out. New bunks, mattresses, etc. required to make the compartment usable.	$15,000
Landing Ship Dock—COLLISION 20 foot gash in ship's side, three vertical frames pushed in, degaussing cables grounded.	$15,000
Guided Missile Frigate—FLOODING After Engineroom flooded to two feet above the lower deck plates, immersing seven pumps and two spring bearings.	$3,000
Fleet Oiler—FIRE Class "B" fire in main engineroom damaged #2 S/S generator, numerous electrical cable runs with some interior structural damage.	$170,000
Cable Repairing Ship—FLOODING After Engineroom flooded, immersing #2 Main Motor, grounding motor and stator windings.	(actual cost) $248,000

These are just a few of the damage-control situations that all too frequently arise in peacetime fleet operations. Fire, flooding, collision, grounding, and heavy weather are constant hazards which the shipboard damage-control organization must be able to cope with rapidly and effectively. This can only be done through proper training.

DAMAGE CONTROL TRAINING

In damage control there is no substitute for realism in training. Experience gained by simulating the conditions that will be encountered in a real emergency can prove to be of great value if ever needed. For this reason, as many of the crew as possible should receive training at one of the Navy's fire-fighting schools. Exposure to the smoke, heat, and noise of an actual fire is one of the most worthwhile advantages of this kind of training. With regard to control of flooding, pipepatching, shoring, and dewatering, it is also imperative that conditions be as realistic as possible. A locally produced training mock-up which provides piping with water under pressure is certainly worth the effort of construction when compared to the experience gained.

Furthermore, such live drills make damage control interesting. Anyone can sit down and read a book and become an authority on just about anything, but damage control tends to be a very dry subject which can lose the student quickly unless some kind of practical application is provided. When the crew is at general quarters, try blindfolding a member of the repair party and instructing him to find his way to a particular location in his party's area of responsibility. The results will be both amusing and educational. Don't allow repair party personnel to lie idly by in the passageways during a gun shoot or missile exercise. Work at having something constructive and informative for them to do. There are no limits to the imagination.

One point that is often overlooked is the value of using moulages to simulate typical battle wounds. More than one strong man has paled at the sight of a nasty-looking chest wound or a severed artery—artificial, of course, but nevertheless effective. Such sights are never pleasant, but damage control is not a pleasant business, and the men need to know this ahead of time.

Items such as oxygen breathing apparatus (OBA), gas masks, and special or protective clothing should be broken out and used frequently. The men need to become familiar with the trappings of their trade so that using them in an emergency will seem natural.

In summary, a good damage-control organization comes only as a result of plenty of study, hard work, attention to detail, and team spirit mixed with a very large measure of imagination. NavShips *Technical Manual,* Chapter 9880, Simulating Damage for Training Purposes, puts it this way:

> Training in making battle repairs in ships is generally limited by circumstances. Occasionally, the need arises to repair a leaky pipe or to renew a small cable; but there is seldom a chance for the average member of a repair party to do any real shoring, to stop a leak in the hull, or to gain experience in any part of damage control outside his own specialty. The most imaginative and energetic organization has to pretend that damage has occurred. There is no way to know if the simulated repairs made would be effective under the pressure and vibration incident to battle conditions. The test comes when the actual damage is sustained.

CHAPTER 14

Specimen Questions for the Shipboard Engineer

The list of questions in this chapter has been prepared to assist the engineer watch officer in a self-evaluation of his own capability. The questions are applicable to the Engineering Department of any ship. They are not intended to be all-inclusive. The watch officer who can answer all of them should have the confidence that he is indeed a shipboard engineer.

Few of these questions are answered in this Guide. As each individual becomes familiar with his shipboard installation, he will acquire the ability to answer them. This familiarity comes with the attainment of a thorough knowledge of the physical location of the various pieces of shipboard power plant equipment, their piping systems, and individual equipment instruction manuals, shipboard publications, and fleet and type commander directives.

BOILERS

Describe the construction of an M-type boiler.

What are the advantages of using water walls in boilers?

What are the causes of slag accumulations in the furnaces of oil-burning, water-tube boilers?

How is a tube plugged?

What are the causes of "panting" during boiler operation?

Trace the path of the steam from the generating tubes to the main and auxiliary lines on a boiler fitted with both a superheater and desuperheater.

Describe and give the purpose of a stud-tube water-cooled furnace wall.

What operating precautions should be taken in connection with economizers?

What are four possible reasons for superheat temperature falling below normal while boiler load remains constant?

What is saturated steam? Superheated steam?

Describe the proper operation of soot blowers.

What causes oil residue on tubes and side walls?

What does white smoke indicate? Black smoke?

What is the function of the dry pipe?

What is the function of a

a. feed-check valve?
b. feed-stop valve?
c. surface blow?
d. bottom blow?

What is flareback in an oil-fired boiler?

What safety precautions must be observed to prevent flarebacks?

What is the cause of priming and foaming in boilers?

What is the function of the boiler safety valve?

What type inspections are required for boilers?

Describe the brickwork in a boiler.

How does a feedwater regulator work?

What is meant by "end point for circulation"?

What type tests must be performed after opening firesides before steaming? Watersides?

Why is it necessary to keep to a minimum air leakage into the furnace through boiler casings?

What is the penalty for tampering with safety valves?

What are the causes and means of preventing corrosion on the waterside of boilers?

Which is the most dangerous of the scale-forming salts normally present in feedwater?

What is the most common source of salt-water leakage to boilers?

What dangers are involved in heating fuel oil?

Explain the principles of operation of automatic combustion control devices for boilers.

What does a hardness test of boiler feedwater indicate?

Which are the areas of greatest corrosion usually found in a boiler?

What are the general requirements in regard to fuel-oil service piping?

What causes a safety valve to pop?

How are burners adjusted for maximum efficiency?

Does fuel temperature have any effect on combustion?

What is scale, and why is it undesirable?

STEAM TURBINES

How do you secure a steam turbine?

What are the preparations and precautions involved in admitting steam to a turbine for warming up?

What are the effects of bearing wear?

Explain the nozzle arrangement operation.

Why is small radial clearance critical in a reaction turbine but not in an impulse turbine?

What is it that distinguishes impulse blading from reactor blading?

What is the purpose of the gland-seal system?

If turbine noise or vibration occurs, what action should be taken?

In general, what is the temperature range at which a main turbine bearing should operate?

What is labyrinth packing?

What is the dummy micrometer used for?

What is the purpose of the turbine-guarding valve?

What precautions must be observed when making an inspection of the interior of the turbine?

What maintenance and operation are required in order to ensure proper vacuum in the turbine installation?

What is the purpose of the gland-exhaust condenser?

What attention should be given main turbines when in port? When idle for indefinite periods?

What precautions are necessary when lifting a turbine?

REDUCTION GEAR

Describe the construction of a reduction gear, including the function of each component.

What is scoring or galling of reduction gears?

What is the normal means of lubricating reduction gears?

CONDENSERS, AIR EJECTORS, VACUUM

How does a condenser maintain a vacuum in the system?

What are the probable causes of faulty air-ejector operation?

What precautions should be observed when replacing air-ejector nozzles?

How do you test a main condenser for shell and tube tightness?

What precautions should be taken when opening a condenser for testing?

What is the normal means of returning drains from the air-ejector condensers?

How do you start the main air ejectors?

What are the most common causes of inadequate condenser vacuum?

Why are zinc protectors installed in condenser watersides?

What are the principal functions of the main and auxiliary condensers?

Why must air be removed from the condenser?

PUMPS AND COMPRESSORS

What is the most usual cause for the improper operation of a

a. reciprocating pump?
b. rotary gear pump?
c. centrifugal pump?

What is pump cavitation? How is it affected by suction-head and fluid temperature?

What is the probable cause of capacity reduction in an air compressor?

What is the purpose of a centrifugal pump volute?

What is the purpose of pump wearing rings?

What are the possible reasons why a boiler feed pump might fail to maintain the water level in a boiler?

What causes "groaning" in the steam end of a reciprocating pump?

How does a duplex steam pump operate?

What is the purpose of unloading relief valves on air compressors?

Why will most pumps not handle highly heated water from below the level of pump suction?

How do you start up a feed booster pump, feed pump, and condensate pump?

Why is it dangerous to operate an air compressor in an engineering space without proper inlet air filters?

What causes an air compressor to secure automatically?

HEAT EXCHANGERS

What heating equipment do the condensate and feedwater normally pass through from the condenser to the boiler on a modern turbine-driven ship?

What are the functions of the deaerating feed tank? Describe its operation for deaerating feedwater.

For what purposes other than feedwater-heating is exhaust steam used aboard ship?

PROPELLERS AND SHAFTING

What inspections of propellers and stern bearings are made during dry-docking?

What type propeller is normally installed? Why?

Give a complete description of the propeller shafting, bearings, etc.

What is the function of the stern tube? How is it constructed? How is it lubricated?

How are spring bearings aligned? What is their purpose?

Describe the stern-tube packing gland.

VALVES

What is the main difference between safety valves and relief valves?

What is the first precaution when renewing a gasket in pressure piping?

What are the advantages and disadvantages of the gate valve in comparison to the globe valve?

What is meant by "bilge injection"?

How do steam separators work?

Why should a gate valve not be used for throttling?

What causes reducing valves to malfunction?

How does a steam trap function?

STEERING AND MISCELLANEOUS MACHINERY

How does an electrohydraulic steering system work?

What type items are to be checked for inspecting an electrohydraulic system?

INSTRUMENTATION

What is the purpose of the manometer?

What is the pyrometer?

Briefly describe and state purpose of

a. tachometers

b. steam calorimeters
c. pneumercators
d. torsion meters.

How does an electrical salinity indicator work?

What is meant by absolute zero temperature?

What is gage pressure in relation to atmospheric pressure and "Hg?

What is a vernier scale? How is it read?

ENGINEERING PRINCIPLES

Describe isothermal and adiabatic expansion.

What is throttling?

What is meant by specific gravity of a substance? Specific heat?

What is latent heat of vaporization of water?

What is a "factor of safety"?

What is meant by P^H value in reference to boiler-water analysis?

What is meant by thermal efficiency of an engine?

Define absolute pressure, atmospheric pressure, acceleration, velocity.

What is meant by back pressure?

What is the difference between a Fahrenheit and a Centigrade thermometer? How do you convert temperatures?

REFRIGERATION

Describe the refrigeration system on your ship.

What is the function of the

a. expansion valve?
b. compressor?
c. condenser?
d. solenoid valve?

e. dehydrator?
f. low-pressure cutout switch?

How do you test for freon leaks?

What causes a compressor to run continuously?

What are the general indications of lack of freon in the system?

ENGINEERING SAFETY

Describe the fixed CO_2 system, steam-smothering system, and automatic sprinkling system.

What should be used in fighting Class A fires? Class B? Class C?

How is an oxygen breathing apparatus operated? Why is it used?

How do you gas-free a tank or void?

Why is ballasting important?

How is a flame safety lamp used? Why?

What are a fog nozzle and its applicator used for?

How are the portable fire pumps operated?

Describe foam systems and their use.

What methods are available for dewatering aboard ship?

LUBRICATING OIL SYSTEM

Describe the lubricating oil system on your ship.

How does a lub-oil purifier function?

What is the purpose of the settling tank?

How does the centrifuge work in removing dirt, water, etc., from the oil?

How long can lubricating oil be used without change?

What causes oil foaming?

ELECTRICITY

How is electricity generated on a steam turbo-generator? AC? DC?

What causes generator brushes to spark?

What is a synchronous motor?

Describe how a transformer works.

What is the importance of meggar readings?

What procedure is used in paralleling two AC generators?

Describe a typical electrical switchboard, including metering devices.

Why should portable tools be grounded?

What is a thermocouple? What is its use?

How is emergency power provided aboard ship?

What safety precautions are necessary for switchboards?

How does a diesel generator operate?

What causes a grooved commutator?

DIESEL ENGINES

How does a diesel engine operate?

What conditions will cause formation of carbon monoxide in the exhaust?

What four major factors control combustion?

Why are cooling systems used on diesel engines?

Describe the diesel cycle.

What is the supercharger used for?

STORAGE BATTERIES

What routine maintenance should be done? How are batteries rated?

How do you treat acid burns?

How do you test battery cells?

What are the indications of a fully charged battery?

What is the difference between an alkaline and an acid battery?

DISTILLING PLANTS

Describe the operation of the vapor compressor, soloshell, or triple-effect evaporator, whichever is installed in your ship.

ORGANIZATION AND ADMINISTRATION

Describe the typical organization of a shipboard Engineering Department. What are the duties of the various division officers?

What are the legal records maintained by the Engineering Department?

What is important about properly maintained engineering logs? How can a good engineer make use of them?

What operating records should be maintained for the Engineering Department?

Why is the accuracy of the daily fuel and water report important?

Write a sample Engineer's Night Order.

Write a sample Monthly Summary.

Discuss your duties in preparing for material and administrative inspections.

Describe the PMS system. Why is it important?

Describe the MDC system. What does it do for your ship?

How can you best be prepared for shipyard overhaul?

Why are proper relations with the shipyard important?

Why are ship trials important? Name five types of trials.

What are the various types of shipboard maintenance?

How are shipalts authorized?

What is important about conducting shipboard training exercises?

Why is ORI important?

APPENDIX A

Boiler Care and Maintenance

1. Inspections, Tests, and Boiler-Cleaning Requirements
2. Records and Reports
3. Safety Precautions

1. **Summary of Inspections, Tests, and Boiler-Cleaning Requirements**

Subject	When required	Reference
Inspect and clean firesides	Each boiler cleaning period after 600-hours steaming or oftener as conditions warrant.	Arts. 9510.42, 9510.43
Tubes	do	Art. 9510.43
Burners	do	Art. 9510.43
Baffles, seal plates, etc.	do	Arts. 9510.43, 9510.166
Casings	During boiler overhauls	Art. 9510.43, 9510.142, 9510.144
Burner opening rings and refractories.	During boiler shut down periods.	Art. 9510.43, 9510.114, 9510.119, 9510.274
Inspect and lubricate sliding feet	At each 600-hours steaming (fireside cleaning) interval.	Art. 9510.28
Inspect metal of headers of header type boilers.	At least once each year	Art. 9510.62
Tramming headers in header type boilers.	Once each year	Art. 9510.64
Inspect and clean watersides.	Each 1,800–2,000-hours steaming or oftener as conditions warrant.	Art. 9150.83
Mechanical cleaning of:		
a. Generating tubes	do	Art. 9510.83
b. Superheater tubes	Each 3,600-4,000-hours steaming or oftener as conditions warrant.	Art. 9510.81 and 9510.82
c. Economizer tubes	Each 1,800-2,000-hours steaming or oftener as conditions warrant.	Art. 9510.81 and 9510.82
Check steam drum water level gage on steaming boiler by blowing down gages	At end of each watch or when there is doubt as to water level in boiler.	Art. 9510.522

Subject	When required	Reference
Hydrostatic tests:		
a. For tightness	After each boiler overhaul or repair not involving renewal of pressure parts and when considered necessary to inspect for leaks	Art. 9510.204
b. After renewal of any pressure parts.	As warranted	Art. 9510.206
c. For strength	Every 5 years if conditions warrant	Art. 9510.206
Steam test of safety valves	Every 1800 operating hours, subsequent to each general ship's overhaul, and after each hydrostatic test	Arts. 9510.211, 9510.207, 9510.302, 9510.312
Hand-lifting test of safety valve easing gear	Every 3 months while raising steam	Art. 9510.210
Casing tightness tests	During naval shipyard overhauls and in any case at least once every 2 years for single-cased boilers and for outer casing of double-cased boilers and at least once every 5 years for inner casing of double-cased boilers.	Arts. 9510.210 and 9510.208
Test operation air casing relief valves	At each 1,800-2,000-hours steaming boiler cleaning period.	Art. 9510.146
Inspect burner drip pans and drain holes on steaming boiler.	Once each watch	Art. 9510.275
Inspect and test steam smothering system.	At each 600-hours steaming (fireside cleaning) interval.	Arts. 9510.582 and 9510.652
Replace burner hose assemblies	Every other shipyard overhaul	Art. 9510.258

2. Records and Reports

a. Data to be Recorded

Subject	Where recorded	Reference
Tube renewals	Boiler tube renewal sheet	Art. 9510.176
Results of inspection of boiler watersides	Engineering log	Art. 9510.83
Method of cleaning boiler watersides	Engineering log	Art. 9510.83
Results of tramming headers	Machinery history cards	Art. 9510.64
Results of hydrostatic tests	Machinery history cards	Arts. 9510.204, 9510.205, 9510.206, 9510.681
Condition of exterior of drums and headers	Machinery history cards	Art. 9510.49
General condition of boiler as determined during cleaning and overhaul	Machinery history cards Engineering log	Arts. 9510.11, 9510.192
Changes in safety valve setting	Machinery history cards Engineering log	Art. 9510.313

b. Reports to be Submitted to Philadelphia Naval Shipyard (Boiler and Turbine Laboratory), Philadelphia, Pennsylvania

Subject	When required	Copies to	Reference
Tube failures			
a. In connection with scale samples	When submitting samples	NAVSHIPS	Arts. 9510.192, 9510.193
b. In connection with water samples	When submitting samples	NAVSHIPS	Arts. 9510.192, 9510.193

c. Items to be Referred to Naval Ship Systems Command

Item	Reference
For approval prior to:	
a. Performing welding on boiler pressure parts	Arts. 9510.221, 9510.222, 9510.223.
b. To establish "Normal Water Level" in steam drum if it cannot be established from plans.	Art. 9510.331.

From NavShips *Technical Manual*, Chapter 9510 (1969 Edition)

3. Safety Precautions

a. *General Safety Precautions*

1. Do not permit men to work in a dead fireroom unless adequate ventilation is provided. Take special precautions to ensure an adequate supply of fresh air in a dead fireroom if boilers in that fireroom are connected to the same smokepipe as steaming boilers in another fireroom.

2. Keep grating over engineroom and fireroom hatches in place. Ensure that all drains and vents to all drums and headers are open before loosening manhole or handhole plates. Stand clear of such fittings when initially opening them after service.

3. Do not stow on engineroom and fireroom gratings articles that obstruct ventilation or that may fall below.

4. Test the entire electric installation in spaces containing flammable vapors for grounds and remedy defects before sending men to work in the vicinity thereof. This test for grounds should be made from a switchboard outside the fouled spaces and repairs made with the circuit dead.

5. Allow no open lights, such as oil lanterns, open flames, candles, or matches, in oil tanks or near oil hose and oil vents.

6. Wipe up any spilled oil at once.

7. Do not allow oil to accumulate on furnace bottoms, in inner casings, bilges, and pockets.

8. Inspect and clean, if necessary, the drain holes and drip pans of registers of air-encased and bulkhead-enclosed boilers at least once each watch.

9. With nonair-encased oil-burning boilers operating in open firerooms, wash or steam out bilges once each watch. With boilers operating under forced draft in closed firerooms (or boiler rooms), inspect the bilges twice each day and clean if necessary.

10. Keep all joints on oil lines tight.

11. Allow smoking in oil-burning firerooms on floor plates in front of steaming boilers only under conditions specified in NavShips *Technical Manual,* Chapter 9510.

12. Keep fire extinguishers in good working condition.

13. Examine and test the steam-smothering system in boiler air casing after each 600 steaming hours during fireside cleaning. Ensure that all pipe perforations are clear and not blocked by loose scale. Cleaning can be effected by clearing each hole with a nail, then removing the cap on each run of pipe and blowing through with steam.

14. Never allow the temperature of fuel oil to be raised to or above the flash point in any part of the system except between the heaters and the atomizers. In no case should oil be heated to a temperature higher than that necessary to reduce the viscosity of the oil to 135 seconds Saybolt Universal.

15. Where the fuel-oil burning system is designed for 300 psi, do not allow oil pressure in any part of the system to exceed that necessary to produce 300 psi at the burner manifolds on the boiler fronts: also, pressure at the fuel-oil service-pump discharge should not exceed 350 psi gage. Where the system is designed for less than 300 psi, such designed pressure should not be exceeded in any point in the system.

16. Do not allow boilers known to have deposits on their heating surfaces, or grease, oil, or other foreign matter in the water to be steamed at high firing rates except in case of great emergency. Such boilers should be cleaned and boiled out at the first available opportunity.

17. Test the straight portions of tubes in water-tube boilers with a straightedge regularly to detect any deflection.

18. Test the boiler steam gages at regular intervals.

19. Ensure that lighting-off torch pot does not leak, and remove all oil from torch pots after boilers are secured and fireroom space is on cold iron or otherwise unmanned.

20. Make no attempt to assist seating of handhole or manhole plates during hydrostatic test until the pressure has been pumped up to 50 psi less than the test pressure to be applied.

b. *Precautions Prior to Placing Boiler in Service*

1. Adhere strictly to prescribed tests of safety valves.

2. Ascertain that all automatic safety features are tested and work properly.

3. In lighting-off, run water just out of sight in water glass and bring water level in sight with auxiliary feed pump. Pump to level about 1 inch above the bottom of the glass of the 10-inch water gage with auxiliary feed pump.

4. Blow through furnaces of oil-burning boiler with air, before lighting-off and before relighting, when all atomizers have been extinguished accidentally or otherwise, except on the superheater side of the superheat control boilers.

5. Use a torch and stand well clear when lighting-off a dead boiler. A torch should likewise be used in lighting-off additional burners if the furnace is relatively cold, or if burners are widely spaced.

6. In boilers with integral superheaters, on or before lighting the first burner, open the superheater discharge connection to auxiliary exhaust line, if so fitted, and open the protection steam connection to the superheater inlet.

7. With a superheat control boiler, never light a burner on the superheater side until adequate steam flow has been established through the superheater. Also, never light a burner on the superheater side unless one or more burners are in operation on the saturated side.

c. *Precautions During Operation of Boiler*

1. Do not exceed authorized maximum steam drum pressure.

2. Never relight atomizers from a hot brick wall.

3. Never leave disconnected atomizers in place.

4. Never close dampers of oil-burning boilers, if installed, except when air-controlled registers are under repair.

5. Test the drain from oil heater at least once each hour.

6. Pump overboard the water in settling tank before using oil from the tank.

7. Do not use oil from a tank in which there is a considerable amount of water mixed with the oil.

8. If suction of the fuel-oil service pump is lost, close the boiler stops before the steam pressure reaches 85 percent of the authorized working pressure of the boiler.

9. In shutting down a boiler, do not close master oil valve until pump is secured, except in an emergency.

10. As long as a boiler is furnishing steam, never shut off the feed supply even for a short period. The water tender assigned to the feed checks should have no other duty than maintaining proper water level.

11. Always remember that a drop in steam pressure without some apparent reason, is a possible indication of low water.

12. Use try cock, if installed, immediately if water falls out of sight in gage glass.

13. Blow through water gages at end of each watch and when there is any question regarding level of water in the boiler.

14. If water drops out of sight in water gages, shut off oil supply or kill the fires, ease open safety valves, close all openings into boiler.

15. Cut out boiler at once, if practical, whenever a brick drops out of the furnace wall.

16. Cut out boiler at once, if practical, whenever oil is discovered or suspected on watersides.

17. During gunfire, see that personnel, as far as practical, stand clear of the possible path of a flareback.

18. Observe the following precautions to reduce danger of flarebacks:

    a. Do not allow oil to accumulate in furnace. Keep atomizer valves tight.
    b. When atomizers are accidentally extinguished, shut off oil and blow through with steam or air before relighting atomizers.
    c. Never attempt to relight an atomizer from a hot brick wall. Use a torch.
    d. When lighting-off with a torch, stand well clear of register to avoid injury in case of flareback.
    e. Avoid white smoke, as it always indicates usage of large amounts of excess air.

19. Never empty a boiler with bottom blow except in emergency.

20. Never blow down the division wall headers, waterscreen headers and water headers of a boiler until after all burners have been secured.

21. When securing a boiler with an integral superheater, open the discharge connection to the auxiliary exhaust before closing the main steam and auxiliary steam-stop valves.

22. When securing a superheat control boiler, secure the burners on the superheater side prior to securing those on the saturated side. Reduce steam temperature slowly as a precaution in avoiding leakage of steam-line joints.

d. *Precautions after Securing Boiler*

1. Remove atomizers from registers as soon as possible after securing.

2. Close all openings to furnaces as soon as all atomizers have been extinguished.

3. Pump water level to three-fourths of a glass when cutting out a boiler.

4. Before removing any fittings or parts subject to pressure, or loosening a manhole or handhole plate fitting of a boiler, take steps to ensure complete absence of pressure or hot water by opening vents and drains to the drums and headers, including the superheater headers.

e. *Precautions in Cleaning Boiler*

1. To prevent accidental opening of valves, close and secure by locking or wiring all connecting valves to boiler when men are working in the boiler.

2. There is always danger that steam and/or hot water may leak to an idle or open boiler through a leaky bottom blow valve when bottom blow valve of another boiler connected to the common blow line is opened. This same danger exists with respect to stop valves, feed valves, and superheater high-pressure drain valves. Open the super-

heater bilge drain valves to permit drainage of any water leaking into headers. If pressure is to be applied to any valve on an open boiler, do not allow any personnel in the boiler until pressure has been applied to the valve and its tightness is positively assured. Take special precautions in checking the tightness of boiler blow and guarding valves of an open boiler which has a common boiler blow connection with a steaming boiler. To assure tightness of the valves on the idle (open) boiler, blow down the steaming boiler before permitting personnel to work on the idle boiler in order to check that the valves of the idle boiler are proven closed and tight. At other times, the boiler blow valves on the commonly connected boilers, both the steaming boiler and opened boiler shall be wired shut and tagged "DO NOT OPEN WITHOUT PERMISSION OF THE ENGINEER OFFICER." When permission is obtained to blow down the steaming boiler, evacuate personnel working in the open boiler until the blowing-down operation is completed. The boiler blow valves of the steaming boiler should then be rewired and tagged.

3. Lash in closed position the control valves of steam-smothering systems while men are working in the vicinity thereof, and remove lashings when the work is finished.

4. Never use naked lights in an open boiler. Insulate thoroughly electric leads of portable lights and use portable lighting fixtures of the watertight type. Hand electric flashlights are preferable. Observe the same precautions when using portable electric lights in the presence of fuel oil, gasoline, kerosene, and other flammable vapors and gases: that is, inspect for and renew defective insulation before use. Remove flammable vapors by ventilation before performing work.

5. Before sending men into a boiler after boiling out, ventilate the boiler thoroughly.

6. While men are working in the interior of a boiler, station a man outside to render assistance as necessary in case of accident.

7. Observe the following precautions in applying metal-conditioning compound to the firesides of a boiler:
   a. Prohibit smoking while the compound is being applied.
   b. Do not apply compound to a boiler if another boiler in the same fireroom is steaming or if the boiler is

connected to the same smoke pipe as another boiler which is steaming.

c. Avoid use of excessive amounts of compound which can collect in pockets subject to high temperature when steaming.

d. Take particular care to detect fires that might occur when boilers are first steamed after application of compound.

e. After a boiler has been sprayed with compound, limit work inside the boiler or uptakes to emergency work only, until the compound has been removed by firing the boiler.

8. Do not close a boiler unless the watersides have been carefully examined for dirt and other foreign matter.

9. Before closing a boiler, inspect the interior to see that no person is left inside and that the boiler, including economizer and superheater headers, is free from tools, loose material, and foreign matter.

From NavShips *Technical Manual,* Chapter 9510 (1969 Edition).

APPENDIX B

Sample Forms

1. Sample Engineroom Warming-up Schedule, Desgen Form 9410-10
2. Sample Fireroom Lighting-off Schedule, ComCruDesPac Gen-9510/21
3. Sample Auxiliary Plant Starting and Operation Check-off List

DESGEN Form 9410–10 (3/55)

ENGINEROOM WARMING-UP SCHEDULE

DD445 and 692 CLASSES

U.S.S.	Date

The following schedule shall be followed each time the main engines are warmed up, unless otherwise directed by the Engineer Officer. It is a check-off list of important steps which must be performed at the scheduled times if the main propulsion plant is to be warmed up properly, and reported ready at ZERO time.

COMMENCE WARMING-UP AT	UNDERWAY AT	STANDARD SPEED WILL BE	SIGNATURE (Engineer Officer)

TIME Before Testing Out			OPERATION
SCHEDULE Hr. Min.	Clock	When Done	
01 - 45			Station engineroom steaming watch. Take and record counter reading and turbine clearances. Enter in Eng. Log specifying cold readings. Prepare Operating Records and Bell Book Sheets.
01 - 35			E. M. energize all necessary power and I. C. circuits.
			Crack all funnel drains. Cross-connect Turbo, Auxiliary steam and exhaust condensate, salt water cooling and H.P. drain systems.
			When line pressures reach 50 p.s.i. cut in H.P. drain traps and close funnel drains.
			Open main condenser overboard and injection valves, and salt water chest vents.
			Open all main propulsion turbine casing and chest drains.
			Back off all throttles, throttle by-pass valves and nozzle valves. Reseat lightly. Test engine "wrong direction alarm."
			Open bulkhead stop valves, cross-over valves and line stop valves for boilers to be cut in on main steam line. Fireroom crack by-pass on steaming boiler's main steam stop valve.
01 - 30			Clean lubricating oil strainers.
			Check oil levels in main sumps, pumps and spring bearings.
			Check temperature of main lube oil sump tanks. If below 90° F. heat oil by connecting steam hose to lube oil cooler from auxiliary exhaust line. Insure that water side of cooler is full of water and that the after vent is open.
			Warm up and start a main lube oil pump (Underway standby pump). Check: oil delivery to all bearings and entire system for leaks.
			Energize and test low pressure lubricating oil alarm operation and setting.
01 - 15			Inspect shaft alleys and adjust glands for proper leak-off.
			Check cooling main and insure adequate cooling water supply to all auxiliaries (25 p.s.i.).
			Engage jacking gear and start jacking main engines. Display a suitable sign at main throttle station.
01 - 10			Warm-up and start a main condensate pump and recirculate condensate through air ejector condenser back to main condenser.
01 - 00			Cut in steam to main turbine glands, check gland leak-offs to insure they are properly lined up, and start gland exhauster fan.
			Line up and cut in steam slowly to 2nd stage of main air ejector.
			Start main condenser circulating water pump. Do not operate with less than 125-150 p.s.i. steam chest pressure.
			Engineroom with hot DFT: Warm up and start main feed booster pump. Warm up main feed pump when suction pressure reaches 40 p.s.i.
			Engineroom with idle DFT: Warm up main feed booster pump and open wide large (2") recirculating valve to DFT.
			Warm up DFT by opening H.P. drains and vents. Crack in auxiliary exhaust until temperature reaches 240° F. Then open wide.
00 - 40			E.M. energize and test the engine-order telegraph and engine-revolution indicator systems, verifying via telephone each position.
			Cut in steam to whistle slowly. By-pass traps until lines are thoroughly warmed up.
			When main condenser vacuum reaches 20", shift auxiliary exhaust and drains to main condenser.

TIME Before Testing Out			OPERATION
SCHEDULE Hr. Min.	Clock	When Done	
00 - 30			Energize and test low pressure lube oil alarm and warm up additional ship's service generator. Connect to line and split the electrical load.
			In conjunction with deck operators, E.M.'s test the anchor windlass and steering-engine. Report results of test to main engine control.
			Bring up main feed pump pressure and commence feeding steaming boiler with main feed.
			Warm up main feed pump (Cold engineroom). Bring up to pressure when ordered to split plant.
00 - 25			Warm up and start lube oil service pump. Adjust standby lube oil pump to deliver 8 p.s.i. to bearings and bring service lube oil pump up to pressure required to cut out standby pump (10 p.s.i. at bearings).
			Warm up and test out standby condensate, booster and main feed pumps.
00 - 20			Cut in 1st stage steam to main air ejector.
			Cut in additional boiler on auxiliary and turbo steam systems. Open boiler main steam stop by-pass valve.
00 - 15			Man the 2JV phones in all main propulsion spaces. Man 1JV phones in enginerooms.
			Notify the Firerooms to open main steam stops and order the plant split. (See Note 2)
			Request permission from the Officer of the Deck to open the guarding valves, disengage the jacking gear and test the main engines.
			Test main engines. (See Note 3)
			Report engineering department's readiness for getting underway to the OOD. (See Note 4)
00 - 00			Request permission from the OOD to spin the main engines every three to five minutes. Ensure that NO way is put on the ship.
			Take and log main turbine axial "hot clearances."
			Standby to answer all bells. (When Command Order is received) Notify all spaces.
			UNDERWAY. Bring main feed pressure up to operating pressure.
00 - 05			Secure main turbine drains.

NOTES

1. The plant shall be fully split at all times when maneuvering in restricted waters.
2. Split the following systems:
   Main steam; Auxiliary; Auxiliary exhaust; High pressure drains; Low pressure drains; Condensate; Booster; Main feed; Reserve feed transfer line; Firemain; Salt water cooling main and the Generator AC and DC loads.
3. When permission is granted to test the main engines:
   Main engine control notify all stations "Standby to test main engines," and After engineroom to "Station shaft observers, disengage jacking gear, crack guarding valve (approx. 5 turns) and report when ready to test engines." Inform throttle stations as to direction to spin main engines but ensure no way is put on the ship.
4. Upon completion of main engine tests, direct all stations to report readiness for getting underway to main engine control and open guarding valves wide open. Main engine control report to OOD that main engines have been tested satisfactorily, engineering department ready to get underway, split plant, on boilers No. and No.
5. The CLOCK times (Column 2) are inserted with the SCHEDULE in Column 1.
6. Submit completed form to the Engineer Officer for his examination and signature prior to reporting readiness for getting underway.

REMARKS: (If any item is unsatisfactory make appropriate comment under remarks.)

EXECUTED (CMM of Watch or MM1 of Watch)	EXAMINED (Engineer Officer)

FIREROOM LIGHTING-OFF SCHEDULE
COMCRUDESPAC-GEN-9510/21 (9-64) *0190-236-2600*

– 600 PSI –

SHIP	DATE

The following shall be utilized in lighting off and cutting in No. boiler. It is a check-off list and the steps must be performed in the prescribed order.

EXPECTED LIGHTING OFF TIME	EXPECTED CUTTING IN TIME	SIGNATURE (Engineer Officer)

TIME BEFORE TESTING OUT HR : MIN	WHEN DONE	OPERATION
		Request permission from OOD to remove stack cover and light off boiler.
		Station steaming watch.
		Cut in auxiliary steam and exhaust lines; insure proper drainage.
		Open all cocks and valves in line to boiler steam gauge, water gauge shut off valves, air cocks, including air vents on the superheater and superheater drains to bilges.
		Drain boiler to bottom of glass if boiler has been idle. Raise the boiler water level to about one inch by opening main feed check, stop, and automatic feed regulating valves to insure that valves are functioning properly.
		Warm up and start one forced draft blower. Ensure that flaps on idle blowers are closed to prevent damage due to lack of lubrication caused by reverse rotation. Blow out and ventilate furnace thoroughly.
		Inspect bottoms and inner fronts of air casings for oil accumulations and clean out any noted prior to lighting off. Make certain that innercasing drain holes are open and clear for conducting any atomizer drip to the furnace.
		Inspect and clean all fuel oil strainers.
		Close individual atomizer root and needle valves.
		Line up fuel oil suction and discharge lines.
		Insert lighting off atomizer assembly. Test operation of air registers.
		Warm up and start one fuel oil service pump (or electric pump).
		Cut in fuel oil heater and circulate oil through heater and burner manifold until oil reaches proper temperature (about 135°–140° F.). By-pass meter during recirculating.
		Examine and operate hand gear for lifting safety valves so far as can be done without lifting valves.
		Using deck control gear, ease up the main, turbo and auxiliary steam stop-valve stems without lifting valve discs off seats. (Where double valve protection is installed, open valve next to boiler first. Then handle second valve with by-pass.)
		E.M. energize and test superheat thermal alarm.
		Close all air registers except No. 1.
		Cut in fuel oil meter.
		Position and light off No. 1 atomizer using 5015 sprayer plate. Always use a torch and stand well clear to avoid a flareback. As soon as steam is formed, shift to No. 2 burner. Thereafter rotate firing from each of the four burners at 10 minute intervals to assure even heat distribution and expansion and resort to intermittent firing as necessary to ensure against a too-rapid rise in boiler steam pressure, especially in the 50 to 250 psi range.
		As boiler heats up and steam forms, frequently check water level for rise due to expansion.
		Close all air cocks and vents after steam has been formed and all air expelled.
		Check boiler gauge to see that it registers pressure.
		Superheater H.P. drains shall be cut in and bilge drains secured when boiler pressure reaches 50-75 lbs. Occasionally open drains to bilges to check drainage of superheater.
		Check operation of gauge glasses by blowing them through.
		Start warming up steam lines using by-passes and opening drains.
		Warm up standby fuel oil service pump.

TIME BEFORE TESTING OUT		OPERATION
HR : MIN	WHEN DONE	
		Warm up and recirculate emergency feed pump on cold suction.
		With boiler pressure slightly above line pressure, cut in on auxiliary, turbo-generator (if applicable), and main steam lines when ordered by the control engineroom. Crack valves slowly, permitting gradual warming up, then open fully.
		Commence feeding boiler. (Directly upon cutting in boiler on auxiliary steam line.)
		Place a set of maneuvering size atomizer assemblies in all idle burners (retracted position).
		Warm up and start additional forced draft blower.
		Report ready to main engineroom.
		Upon completion of main engine test report readiness for getting underway to main engine control. Take fuel oil meter reading.
		Light fires under superheaters upon orders from main engine control.

NOTES

1. Whenever possible, stack covers shall be removed during daylight hours. Secure all electric power lines with insulators connected to the stack, prior to going aloft. Wear safety belts.
2. When all steam generated in the boiler is taken from the auxiliary steam connection and there is no steam flow through the superheater, a firing rate of approximately 136 gal. per hour must not be exceeded, excepting in an emergency, to prevent overheating and damaging the superheater.
3. At no time while raising steam in a boiler and prior to cutting it in on the main steam line shall a burner be lighted under the superheater side except on the specific orders of the Chief Engineer. No superheater burners should be lighted unless the superheater protection device indicates a safe steam flow through the superheater. A minimum safe limit is 2 inches of water as indicated by the low flow indicator.
4. The superheat temperature should be raised evenly on all boilers at a rate not to exceed 10 degrees F. per minute.

REMARKS

THE BTC OR BT IN CHARGE OF THE LIGHTING OFF WATCH WILL SIGN THIS SHEET AND TURN IN WITH OTHER FIREROOM LOGS.

SIGNATURE *(BT in charge)*	SIGNATURE *(Engineer Officer)*

COMCRUDESPAC ENGINEERING SCHOOL

Building 78, U. S. Naval Station Bln 160
San Diego, California 92136 EO-6

AUXILIARY PLANT STARTING AND OPERATION

1. Preliminary steps taken to place the auxiliary plant in operation include the following:
   a. Close the header drain valves and open the header vent valves.
   b. Open the auxiliary circulating pump sea suction.
   c. Open the auxiliary circulating water overboard valve.
   d. Start the auxiliary circulating pump.
2. Before the auxiliary condensate pump may be started, it will be necessary to establish a water level in the auxiliary condenser.
3. The following methods may be employed to establish a water level in the auxiliary condenser hot well:
   a. Run water down the DA tank into the make up feed line. Be sure the make up feed valves on the feed tanks are closed.
   b. If the feed tank level is high, it may be possible to siphon water from that tank.
   c. From main condenser through recirculating piping.
4. Prior to placing the DA tank in service, it will be necessary to bring the tank level to the operating level. Water may be admitted to the DA tank from the following sources:
   a. Main or auxiliary condensers.
   b. From other engineroom through condensate crossover.
   c. Feed tank with the emergency feed pump.
5. Open the discharge valve to the DA tank and open the recirculating valve to the auxiliary condenser.
6. Line up and start the auxiliary condensate pump.
7. After the auxiliary condensate pump has been started the auxiliary air ejector may be placed in operation.
8. When the vacuum has reached about 15″ it is permissible to cut in the auxiliary exhaust and the low pressure drains.
9. If the DA tank is to be used, it should be heating while vacuum is being raised on the auxiliary condenser.
   a. Start a booster pump recirculating to the DA tank through the 2″ recirculating valve.
   b. Crack open the auxiliary exhaust inlet valve.

c. The high pressure drains may be cracked in to help warm up the DA tank.
d. The DA tank operating temperature should be attained in approximately 20 minutes. Continue to circulate about 10 minutes with the auxiliary exhaust and LP drain inlet valves wide open to deaerate the water.
e. Prior to reporting a hot suction available for the emergency feed pump, it should be established that a tank is lined up for make up feed, water in the DA tank has reached the proper temperature and has been deaerated, and also that the 2″ recirculating valve is closed and the ¾″ valve is open.

10. The turbo-generator may be placed in operation when vacuum in the auxiliary condenser has reached 15 inches.

11. If the temperature of the condensate leaving the air ejector exceeds 140 degrees it will be necessary to recirculate with the hand control valve.

12. Cause for loss of vacuum in the auxiliary plant will be approximately the same as in the main plant:
    a. Inadequate heat exchange from the salt and fresh water sides.
    b. Failure of the air ejectors.
    c. Inadequate condensate removal.
    d. Air leak into the system.

SECURING THE AUXILIARY PLANT

1. Secure generator if running.
2. Secure auxiliary exhaust and LP drains to condenser.
3. Continue operation of air ejector for 15 minutes.
4. Secure root steam to air ejector.
5. Circulate condensate through the air ejector condenser until cool, then secure the auxiliary condensate pump.
6. When the auxiliary condenser is cooled, secure the auxiliary circulating pump.
7. Close injection and overboard valves.
8. Close turbo-exhaust valve.
9. Close condensate and air ejector valves.

APPENDIX C

Summary of Feedwater Tests

1. Limits on Feedwater Components and Frequency of Test
2. Frequency of Sampling Boiler Feedwater
3. Mandatory Tests on Boiler Feedwater Samples
4. Boiler Feedwater Limits and Controls

1. Limits on Feedwater Components and Frequency of Test

Feedwater Component	Limit	Test Frequency
Evaporator distillate		
a. Measuring tank	Chloride-0.065 epm max.	Each filling
Make-up feedwater		
a. Reserve feed tanks	Chloride-0.25 epm max.	Daily and prior to use
	Hardness-0.20 epm max.	Daily and prior to use
Condensate		
a. Main condenser	Chloride-0.05 epm max.	Every 15 minutes underway Every 30 minutes standby
b. Auxiliary	Chloride 0.05 epm max.	Every 30 minutes
Feedwater		
a. Deaerating feed tank	Chloride-0.15 epm max.	Each watch
	Hardness-0.20 epm max.	Daily
	Dissolved Oxygen-0.02 ppm max.	Daily
b. Surge tanks and/or other feed tanks on line	Chloride-0.15 epm max.	Each watch
	Hardness-0.20 epm max.	Daily

2. Frequency of Sampling Boiler Feedwater

Freshly filled: 600-psi boilers, immediately; 1200-psi and diatomite filter ships' boilers, hourly until proper water conditions are established.

Steaming: Daily and after any chemical treatment or blowdown.

Idle, wet lay-up: Weekly; after any chemical treatment or blowdown; and prior to light-off.

3. Mandatory Tests on Boiler Feedwater Samples

Boiler Pressure

600 psi	1200 psi	Diatomite Filter Ships
Alkalinity	pH	Alkalinity
Hardness	Phosphate	Phosphate
Chloride	Chloride	Chloride
Conductivity	Conductivity	Conductivity

4. Boiler Feedwater Limits and Controls

Boiler	Test	Limit	Control
600 psi & below	Chloride	2.0 emp max.	Blowdown
	Alkalinity	2.5-3.5 emp range	Navy boiler compound and blowdown
	Hardness	0 epm	Navy boiler compound
	Conductivity	1300 mmhos max.	Blowdown
1200 psi	Chloride	2.0 epm max.	Blowdown
	pH	10.4-11.0 range	Caustic soda/ blowdown
	Phosphate	10-25 ppm range	Disodium phosphate/ blowdown
	Conductivity	700 mmhos max.	Blowdown
Diatomite Filter Ships	Chloride	2.0 epm max.	Blowdown
	Alkalinity	2.5-3.5 epm range	Caustic soda/ blowdown
	Phosphate	20-100 ppm range	Disodium phosphate/ blowdown
	Conductivity	1300 mmhos max.	Blowdown

NOTES: 1. Each shipment of shore feedwater ("rainmaker" shore steam condensate or commercial) should be accompanied by a certificate of analysis showing that the water meets the following requirements:

pH—5.0-9.0
Dissolved solids—10 ppm max.
Conductivity—25 mmhos/cm max.
Dissolved silica—0.2 ppm max.
Hardness—0.1 epm max.

2. Each shipment of shore feedwater should be sampled and tested prior to acceptance for agreement with the following requirements:

Hardness—0.1 epm max.
Alkalinity—Colorless to phenolphthalein, green to methyl purple.

APPENDIX D

Extracts from DDC Reports

Summary Results Based on Ships Reporting Through the 3-M System

1. AD 838998, High-Pressure Air Compressor Maintenance Analysis, June 1967
2. AD 839103, Forced-Draft Blowers Maintenance Evaluation Analysis Report Problem Identification Program, June 1968
3. AD 839041, Mark 19 Mod 3, 3A, and 3B Gyrocompass Equipment Maintenance Analysis Report, July 1968
4. AD 847630, Distilling Plant, Vapor Compression Maintenance Analysis Report, October 1968

1. AD 838998, High-Pressure Air Compressor Maintenance Analysis, June 1967

   This Ingersoll-Rand Compressor CID Number 061430124, was determined to have a Mean Time Between Corrective Maintenance (MTBCM) of 66 hours and a Mean Time Between Parts Replacement (MTBPR) of 163 hours. The major problem areas were identified as *Cylinder Valves* (MTBPR-340 hours); Lube Oil Pump, Salt Water Pump and Cooler Seals (MTBPR-453 hours); and Piston/Cylinder Assemblies (MTBPR-510 hours). The Lube Oil Pump, Salt Water Pump, and Cooler Seals are considered to be the "worst case" problem in this compressor due to certain maintainability and safety aspects. The application of proper lubrication is recommended along with modification of salt water cooling to fresh water cooling.

2. AD 839103, Forced-Draft Blowers Maintenance Evaluation Analysis Report Problem Identification Program, June 1968

   For the one-year period of this report, approximately one-million operating hours were determined for 584 forced-draft blowers on 63 selected ships. These same units required 59,154 maintenance manhours and $85,465 in parts costs.

   In the comparison of the three selected manufacturers, the overall performance of Hardie-Tynes was the worst for all selected parameters; MTBCM of 600 hours, MMTR of 47.3 hours, MTBPR of 1616 hours, parts cost per unit of $244, and parts cost per 1000 operating hours. Westinghouse achieved the highest in mean time between corrective maintenance (986 hours) and mean time between parts replacement (3943 hours), while Carrier-Elliott achieved the lowest mean man-hours to repair (16.6 manhours). Both Westinghouse and Carrier-Elliott were about equal in terms of parts costs per unit and parts cost per 1000 operating hours.

   For Westinghouse, leakage and excessive wear of packing gland and thrust bearing assemblies and lube oil cooler leaks were the major problem areas.

   For Hardie-Tynes, cut nozzle blocks and reversing chambers and excessive wear in the governor speed limiting assemblies were the major problem areas.

   For Carrier-Elliott, leaking nozzle blocks and reversing chambers and lube oil coolers were the major problem areas.

   "Leaking" was the malfunction most prevalent in all blowers considered in this report. There may be a variety of causes to such a problem. It is beyond the scope of this report to determine causes; within the limited scope it will not be possible to relate malfunctions and causes with design, installation or operation.

The above comparisons are limited to the ships listed in this report and should be used with caution when extrapolating performance of the manufacturer's blowers on other ships.

3. AD 839041, Mark 19 Mod 3, 3A, and 3B Gyrocompass Equipment Maintenance Analysis Report, July 1968

Maintenance data reports on the MARK 19 MOD 3, 3A, and 3B Gyrocompass Equipment submitted by 222 ships during the 12-month period 1 January 1967 through 31 December 1967 are analyzed. It is shown that 304 equipments were responsible for 2110 corrective maintenance jobs requiring 24,178 manhours, an average of 11.5 manhours per job. This amounts to an average of 6.9 jobs and 79.5 manhours per equipment.

The observed Mean Time Between Failures (MTBF) was: 496 hours for the MOD 3, 749 hours for the MOD 3A, and 802 hours for the MOD 3B. The three major component MTBF values are calculated and by comparison the Control Cabinet is shown to most limit the equipment reliability in all three MOD's and the Standby Motor-Generator further reduces the MTBF of the MOD 3 and 3A equipment. The average per equipment parts costs was: $791 for the MOD 3, $514 for the MOD 3A, and $374 for the MOD 3B. The amount of planned maintenance accomplished is shown to be below requirements, with 71 percent (3129 routines) of the required routines reported but expending 177 percent (4596 manhours) of the required estimated manhours.

Motor-Generator type Standby Power Supply parts that were replaced are identified and their replacement rates calculated. The Motor-Generators are considered to have poor maintainability since they consumed an average 20.9 manhours per CM job.

The ship type analysis identifies the MARK 19 MOD 3, 3A, and 3B Gyrocompass Equipments as achieving a higher MTBF in ships considered as stable platforms.

Conclusions are drawn from the analyses that include the following:

(a) The MARK 19 MOD 3B Gyrocompass met its estimated MTBF considering the equipment complexity in terms of active elements.

(b) The MARK 19 MOD 3B and 3A Gyrocompass Equipment achieved a significantly higher MTBF than the MOD 3. The Control Cabinet and Motor-Generator components were most responsible for the MOD 3 Gyrocompass' lower MTBF.

(c) Design improvements in Control Cabinets and Motor-Generators would contribute most toward an increase in MARK 19 Gyrocompass reliability.

4. AD 847630, Distilling Plant, Vapor Compression Maintenance Analysis Report, October 1968

Each ship considered in this report required an average of 4.8 CM actions, 53.5 CM manhours, and $262.45 parts cost per year.

Preheaters accounted for 2.3% of the total CM actions, 0.9% of the total CM manhours, 2.5% of the total parts cost, and 0.3% of the total parts replaced. The reported data indicates that preheaters are not a problem.

Vapor compressors are the most significant problem area within the distilling plants and accounted for 52.8% of the total parts cost (by CID) and 26.1% of the total CM manhours (by EIC). Seal assemblies (FSN 4620-733-5061 and 4310-875-6737) accounted for 93.3% of the compressor parts cost—in most cases, bearings were also replaced. The high CM manhour expenditure resulted from the major disassembly effort required to replace the seals and bearings. Eight of 67 ships replaced seals on eight units during this reporting period.

Immersion heaters were the next high cost item and accounted for 11.5% of the total parts cost. Six ships replaced 13 heaters (FSN 4540-722-5263 and 4540-585-6332) out of a heater population of 252.

Heat exchangers were the highest part replacement equipment and accounted for 24.5% of the total parts replaced and 15.5% of the total CM manhours. Nine ships replaced 553 ferrules in nine of 67 units.

APPENDIX E

ENGINEERING SYMBOLS

PIPE FITTINGS, TYPES OF CONNECTIONS

- SCREWED ENDS
- FLANGED ENDS
- BELL-AND-SPIGOT ENDS
- WELDED AND BRAZED ENDS
- SOLDERED ENDS

ELBOWS

FITTING	SYMBOL
ELBOW, 90 DEGREES	
ELBOW, 45 DEGREES	
ELBOW, OTHER THAN 90 OR 45 DEGREES, SPECIFY ANGLE	30°
ELBOW, LONG RADIUS	LR
ELBOW, REDUCING	
ELBOW, SIDE OUTLET, OUTLET DOWN	
ELBOW, SIDE OUTLET, OUTLET UP	
ELBOW, TURNED DOWN	
ELBOW, TURNED UP	
ELBOW, UNION	

TEES

FITTING	SYMBOL
TEE	
TEE, DOUBLE SWEEP	
TEE, OUTLET DOWN	
TEE, OUTLET UP	
TEE, SINGLE SWEEP, OR PLAIN T-Y	

OTHER PIPE FITTINGS

FITTING	SYMBOL
BUSHING	
CAP	
COUPLING	
PLUG	
REDUCER, CONCENTRIC	
UNION, FLANGED	
UNION, SCREWED	
EXPANSION JOINT, BELLOWS	
EXPANSION JOINT, SLIDING	

VALVES, TYPES OF CONNECTIONS

- SCREWED ENDS
- FLANGED ENDS
- BELL-AND-SPIGOT ENDS
- WELDED AND BRAZED ENDS
- SOLDERED ENDS

STOP VALVES

VALVE	SYMBOL
GENERAL SYMBOL	
ANGLE	
GATE	
GATE, ANGLE	
GLOBE	
GLOBE, AIR OPERATED, SPRING CLOSING	
GLOBE, DECK OPERATED	
GLOBE, HYDRAULICALLY OPERATED	H
STOP COCK, PLUG OR CYLINDER VALVE, 2 WAY	
STOP COCK, PLUG OR CYLINDER VALVE, 3 WAY, 2 PORT	
STOP COCK, PLUG OR CYLINDER VALVE, 3 WAY, 3 PORT	
STOP COCK, PLUG OR CYLINDER VALVE, 4 WAY, 4 PORT	

RELIEF, REGULATING, AND SAFETY VALVES

VALVE	SYMBOL
GENERAL SYMBOL	
ANGLE, RELIEF	
BACK PRESSURE	
GLOBE, RELIEF	
GLOBE, RELIEF ADJUSTABLE, OR SPRING LOADED REDUCING	
PRESSURE REDUCING OR PRESSURE REGULATING, INCREASED ACTUATING PRESSURE CLOSES VALVE	
PRESSURE REDUCING OR PRESSURE REGULATING, INCREASED ACTUATING PRESSURE OPENS VALVE	
PRESSURE REGULATING, WEIGHT-LOADED	
SAFETY, BOILER	

CHECK VALVES

VALVE	SYMBOL
GENERAL SYMBOL	
CHECK, LIFT	
CHECK, SWING	
GLOBE, STOP CHECK	

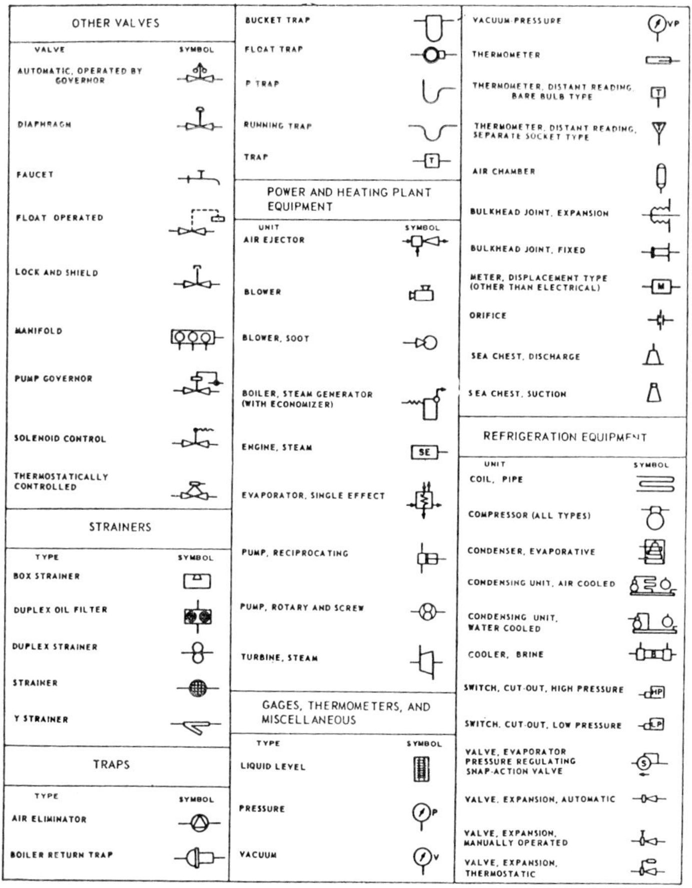

OTHER VALVES	
VALVE	SYMBOL
AUTOMATIC, OPERATED BY GOVERNOR	
DIAPHRAGM	
FAUCET	
FLOAT OPERATED	
LOCK AND SHIELD	
MANIFOLD	
PUMP GOVERNOR	
SOLENOID CONTROL	
THERMOSTATICALLY CONTROLLED	

STRAINERS	
TYPE	SYMBOL
BOX STRAINER	
DUPLEX OIL FILTER	
DUPLEX STRAINER	
STRAINER	
Y STRAINER	

TRAPS	
TYPE	SYMBOL
AIR ELIMINATOR	
BOILER RETURN TRAP	
BUCKET TRAP	
FLOAT TRAP	
P TRAP	
RUNNING TRAP	
TRAP	T

POWER AND HEATING PLANT EQUIPMENT	
UNIT	SYMBOL
AIR EJECTOR	
BLOWER	
BLOWER, SOOT	
BOILER, STEAM GENERATOR (WITH ECONOMIZER)	
ENGINE, STEAM	SE
EVAPORATOR, SINGLE EFFECT	
PUMP, RECIPROCATING	
PUMP, ROTARY AND SCREW	
TURBINE, STEAM	

GAGES, THERMOMETERS, AND MISCELLANEOUS	
TYPE	SYMBOL
LIQUID LEVEL	
PRESSURE	P
VACUUM	V
VACUUM-PRESSURE	VP
THERMOMETER	
THERMOMETER, DISTANT READING, BARE BULB TYPE	T
THERMOMETER, DISTANT READING, SEPARATE SOCKET TYPE	
AIR CHAMBER	
BULKHEAD JOINT, EXPANSION	
BULKHEAD JOINT, FIXED	
METER, DISPLACEMENT TYPE (OTHER THAN ELECTRICAL)	M
ORIFICE	
SEA CHEST, DISCHARGE	
SEA CHEST, SUCTION	

REFRIGERATION EQUIPMENT	
UNIT	SYMBOL
COIL, PIPE	
COMPRESSOR (ALL TYPES)	
CONDENSER, EVAPORATIVE	
CONDENSING UNIT, AIR COOLED	
CONDENSING UNIT, WATER COOLED	
COOLER, BRINE	
SWITCH, CUT-OUT, HIGH PRESSURE	HP
SWITCH, CUT-OUT, LOW PRESSURE	LP
VALVE, EVAPORATOR PRESSURE REGULATING SNAP-ACTION VALVE	S
VALVE, EXPANSION, AUTOMATIC	
VALVE, EXPANSION, MANUALLY OPERATED	
VALVE, EXPANSION, THERMOSTATIC	

NavPers 10788 B, *Principles of Naval Engineering*

ABBREVIATIONS

AD—Destroyer Tender
ADMIN—Administrative Inspection
AFS—Combat Stores Ship
AUX—Auxiliary
BR—Boilermaker
BT—Boiler Technician
BTU—British Thermal Unit
CASREP—Casualty Summary Report
CG—Guided Missile Cruiser
CIC—Combat Information Center
CODAG—Combined Diesel and Gas Turbine
COMCRUDESPAC, CCDP—Commander Cruiser-Destroyer Force, Pacific
COSAG—Combined Steam and Gas Turbine
COSAL—Coordinated Shipboard Allowance List
CRUDESPAC—Cruiser-Destroyer Force, Pacific
CVA—Attack Aircraft Carrier
DCA—Damage Control Assistant
DCC—Damage Control Central
DD—Destroyer
DDC—Defense Documentation Center
DDCPO—Division Damage Control Petty Officer
DDG—Guided Missile Destroyer
DE—Destroyer Escort
DEG—Guided Missile Escort Ship
DFT—Deaerating Feed Tank
DIV—Division
DLG—Guided Missile Destroyer Leader, Guided Missile Frigate
DLGN—Guided Missile Frigate, Nuclear-Powered
EDDO—Engineering Department Duty Officer In Port
ELEC. OFF.—Electrical Officer
EMF—Electromotive Force
EM—Electrician's Mate
EN—Engineman
ENGR. OFF.—Engineer Officer
EOCC—Engineering Operational Casualty Control
EOOW—Engineering Officer of the Watch
EOSS—Engineering Operational Sequencing System
EOT—Engine Order Telegraph
EPM—Equivalent Parts Per Million
EVAP—Evaporator
FA—Fireman Apprentice
FAT—Final Acceptance Trials
FN—Fireman
FTG—Fleet Training Group
FXP—Fleet Exercise Publication
Hg—Mercury
HT—Hull Maintenance Technician
IC—Internal Communications, Internal Communications Electrician
INSURV—Inspection and Survey (Board)
JP-5—Jet Fuel (Navy aircraft)
LAB—Laboratory

LKA—Amphibious Cargo Ship
LP—Low Pressure
LST—Landing Ship, Tank
MAT—Material Inspection
MDC—Maintenance Data Collection
MFP—Main Feed Pump
MI—Material Inspection
ML—Molder
MM—Machinist's Mate
MMHOS—Micromhos
MPA—Main Propulsion Assistant
MR—Machinery Repairman
MSC—Minesweeper, Coastal (non-magnetic)
MSO—Minesweeper, Ocean (non-magnetic)
NAVPERS—Bureau of Naval Personnel
NAVSHIPS—NAVSHIPSYSCOM
NAVISHIPSYSCOM—Naval Ship Systems Command
NBC—Nuclear, Biological, Chemical (warfare) defense
NSFO—Navy Special Fuel Oil
NWIP—Naval Warfare Information Publication
OBA—Oxygen Breathing Apparatus
OOD—Officer of the Deck
OPNAV—Office of the Chief of Naval Operations
ORI—Operational Readiness Inspection
PAT—Preliminary Acceptance Trial
PM—Patternmaker
PMS—Planned Maintenance System
POSDCORB—Planning Organizing Staffing Directing Controlling Observing Reporting and Budgeting
PO—Petty Officer
PQS—Personnel Qualification Standards
PSI—Pounds Per Square Inch
PSIG—Pounds Per Square Inch Gage
PPM—Parts Per Million
RED. VLV.—Reducing Valve
RFS—Ready for Sea
RPM—Revolutions Per Minute
SFR—Specific Fuel Rate
SHP—Shaft Horsepower
SSTG—Ship Service Turbogenerator
SUPSHIP—Supervisor of Shipbuilding
TYCOM—Type Commander
UCMJ—Uniform Code of Military Justice
YN—Yeoman
3-M—Standard Navy Maintenance and Material Management

GLOSSARY

Administration. The management of all phases of naval operations not directly concerned with strategy or tactics.

Air Ejector. A type of jet pump, having no moving parts, which uses high-velocity steam flow to remove air and other noncondensable gases from a condenser in order to maintain vacuum and to reduce corrosion by decreasing oxygen content.

Air Ejector Condenser. Heat exchanger which receives the steam-air mixture from the air-ejector assembly, condenses the steam in the mixture, and returns the condensate to the condensate system while venting the air to the atmosphere.

Air Registers. A subsystem of the boiler designed to regulate air flow into the boiler fire box, evenly mix the air and fuel in a ratio for proper combustion, and prevent flame from being blown back from the fuel-oil atomizer.

Alkalinity. The quality or state of feedwater (or any solution) in which certain impurities bring about an excess of alkali over an acid; normally measured by a pH number, pH7 being the neutral point.

Allowance. Authorized personnel on a peacetime level, reduced from the wartime complement, based on peacetime operations, habitability, budgetary consideratons, and upkeep requirements.

Allowance List. A listing that provides the ship with repair parts, special tools, and other equipment required to be on board to support installed equipment and, in addition, a list of portable and semi-portable items of equipage authorized for shipboard use.

Astern Turbine. The astern elements installed either in the exhaust ends of the low-pressure turbine or in the exhaust end of a single-casing system, which are controlled by an astern throttle for backing or reversing the propulsion shaft.

Auxiliary Boiler. Normally, a ship's boiler that is used only to provide ship-service steam and is not part of the main-propulsion system.

Auxiliary Condenser. Normally, the heat exchanger used to condense turbo-generator exhaust steam and to receive condensate drains from the auxiliary air ejectors.

Auxiliary Exhaust Steam. Steam exhausted from the turbo-generator and from steam-driven auxiliaries, such as pumps and forced-draft blowers, which does not exhaust to the main or auxiliary condensers.

Auxiliary Machinery. All machinery other than the main propulsion boilers, turbines, and condensers.

Auxiliary Steam. Steam, other than that used to drive main-propulsion units, which is required at various temperatures and pressures for operating many systems and units of machinery both inside and outside of the engineering spaces such as fuel oil heaters, whistles, sirens, and a variety of pumps.

Availability. Period assigned a ship for accomplishment of work at a repair facility. May be restricted, technical, regular overhauling, voyage repairs, or upkeep period availability.

Babbitt. A soft, white antifriction alloy of copper, tin, and antimony used for bearing surfaces.

Back Pressure. Pressure on exhaust side of a steam auxiliary or prime mover.

Back Suction. The suction taken by a delivery ship on the fueling hose after fuel has been delivered.

Bell Book, Engineer's. Official legal record of the engine orders received in the engineroom from the bridge.

Bilge. Inside bottom of a ship or boat.

Black Gang. Engineering Department personnel. A carryover from the days when coal was used as the ship's fuel.

Blowdown. The expelling of a portion of the water from the boiler by use of the boiler steam-drum pressure.

Blow Tubes. To inject steam into the firesides of a boiler via the soot-blower system for the purpose of removing soot deposits from the boiler tubes.

Boiler. The source or high-temperature region of the thermodynamic cycle in a conventional steam-turbine propulsion system. Steam generated in the boiler is converted into mechanical energy in main turbines, turbine-driven auxiliaries, reciprocating pumps, etc.

Booklet of General Plans. A book containing plans of a ship, including individual deck, platform, hold, and inboard and outboard profile plans.

Bottom Blowdown. The expelling of a portion of the boiler feedwater from a low unit of the boiler.

Bridge Gauge. An instrument used to measure the vertical and lateral clearances of a journal in any type of sleeve bearing made in halves. So called because it bridges between the two sides of the lower bearing half.

Brine Pump. Any condensate-type pump, usually vertical, that takes a suction on the high-salinity high-density brine fluid in the evaporator and discharges it into an overboard discharge line.

Bulkhead Stop. Any ship-system valve that serves to isolate the system within a major compartment. Normally located at the compartment bulkhead several inches prior to the system's piping bulkhead penetration.

Bus-Ties. Main conductors used to connect one switchboard to another, or to permit the paralleling of the ship-service generators.

By the Numbers. Everything in a prescribed sequence.

Carryover. Water carried out of the boiler with the steam.

Chloride. Sea-salt contamination of water.

Churning. Violent agitation that can cause foaming of lubricating oil.

Circulating Water. Sea water pumped or scooped into the waterside of condensers and used as the heat sink for turning exhaust steam into liquid condensate.

Class A Fire. Fire involving ordinary combustible materials, such as wood, paper, mattresses, clothing, etc.

Class B Fire. Fire involving oils, gasoline, greases, and other liquid inflammables.

Class C Fire. Fire in electrical equipment.

Cold Iron. The state of having no machinery operating in the main engineering spaces.

Compartment Check-Off List. A list of fittings, their location, and function in a compartment. Used for damage-control purposes.

Concentration. The relative content of dissolved material. Sometimes also applied to undissolved (suspended) material in a solution.

Condensate. Exhaust steam that has been cooled and converted back to liquid in the condenser.

Condensate Pump. Pump used to remove condensate from a condenser's hot well and to deliver it to the deaerating feed tank via the air-ejector condensers.

Condition Watches. Various conditions of battle readiness or simulated war operations which do not require full-crew manning, as in the case of general quarters, but do require an increase in stations which are manned over and above normal steaming watches.

Contamination, Boiler. An excessive concentration of sea salts in distillate or any water component of the boiler cycle. Also, any foreign material in the water such as fuel oil, lube oil, corrosion products, etc.

Corrosion, System. Boiler metal being converted to its oxide and/or dissolving in the boiler water. Also, dissolving of the metals in the pre-boiler and after-boiler systems by feedwater and steam, respectively.

Couplings. Connections that provide longitudinal flexibility between the driving and driven shafts, and permit each shaft to be adjusted and held in its proper axial position.

Crown Thickness Measurement. Direct micrometer measurement of the working half of the bearing shell and comparison with original or previous readings to determine bearing wear. Normal procedure for determining wear on reduction gear bearings.

Cruising Range. The endurance of a ship in nautical miles at moderate or cruising speed.

Cruising Speed. Speed established for most economical operation on the cruising turbine combinations (normally, < 20 knots).

Cruising Turbine. A turbine normally connected to the high-pressure turbine rotor through a reduction gear and used to contribute power to the propulsion shaft at speeds up to the most economical, or cruising, speed.

Damage Control Book. Shipboard publication containing information, in the form of texts, tables, and plates, concerning facilities and characteristics of the ship's piping, wiring, liquid loading, and compartmentation systems.

Damage Control Central. Central station located in safest area of ship from which the DCA directs measures for the control of damage and preservation of the ship's fighting capability during general quarters.

Damage Control Posture. Measures necessary to preserve and re-establish watertight integrity, stability, and offensive power; to make rapid material repairs; to control list and trim; to limit spreading of fire and flooding; to protect against toxic agents; and to care for the wounded.

Deaerating Feed Tank. A tank that removes dissolved oxygen from the condensate, heats the condensate and acts as a surge tank before the condensate enters the feed system.

Degaussing System. System designed to reduce the magnetic field of a ship in order to protect against magnetic mines.

Dissolved Oxygen. Air that is dissolved in water.

Dissolved Solids. Sea salts, corrosion products, treatment chemicals, and other soluble material dissolved in boiler water or other waters. Dissolved solids are in solution and cannot be seen.

Distillate. The useful product of the ship's evaporators. It may be considered as highly diluted sea water, since it contains about 3 ppm of sea salts, whereas sea water contains 36,000 ppm of sea salts.

Divided Furnace Boiler. A boiler design that contains a superheater furnace and a saturated side furnace, separated from each other by a waterscreen. Examples are A-type and M-type boilers.

Division. The basic administrative unit into which a ship's crew is organized.

Dock Trials. Test of ship's propulsion machinery while alongside dock after construction or overhaul but before sea trials.

Docking Plan. Plan showing arrangements for installation of blocking in the selection of dry-dock position.

Docking Report. A report submitted by a dry-docking activity concerning major work accomplished on shafts and propellers; bearing clearances; any areas of severe corrosion or wear; last docking position; and date of last dry-docking.

Domino Effect. A chain of events occurring in rapid succession; one event triggering one or more additional events, such as one casualty causing other casualties.

Downtime. Time that a piece of machinery is out of commission or inoperable.

Dump. The complete emptying of a boiler by draining all the feedwater into the bilge.

Economic Speed. The speed at which the best specific fuel rate is obtained.

Economizer. Heat-transfer device in uptake area of boiler which utilizes escaping stack gases to preheat the feedwater, thus reducing the amount of fuel oil required to generate saturated steam in the boiler.

Electric-Drive Turbine. A propulsion system which includes a turbine, a main generator, a propulsion motor, a DC generator for supplying excitation current to the generator and propulsion motor, and a propulsion control board.

Endurance. Time a ship can steam at cruising speed without fuel-oil replenishment.

Engineer Officer. Head of the Engineering Department, in accordance with *Navy Regulations,* 1948.

Engineering Logs. The numerous worksheets, operating records, and accounts that are the basis of a well-administered Engineering Department. Some are mandatory, others are recommended.

Engineering Subspecialist. An unrestricted line officer with specialized training in shipboard engineering.

Evaporator. A device for making fresh water from salt water by evaporation.

Express-Type Boiler. A boiler designed to raise steam quickly and to respond rapidly to demands for steam changes.

Fatigue. Material failure due to a time-varying load, either repeated or alternating; or an alternating load superimposed on a mean load.

Feed Booster Pump. A pump that provides feedwater to the main feed pump because, in most ships, the deaerating feed tank is, by necessity, located in a position from which it cannot provide a positive suction head.

Feed Pump, Main. A pump that converts a high-velocity head into a high-pressure head in order to pump feedwater into the boiler.

Feed Tanks. Storage tanks for feedwater.

Feedwater. The water fed to the boiler. It contains less salt than potable water and it may consist, at any one time, of one or any combination of the following: distillate, make-up feedwater, and condensate.

Field Changes. Mandatory alterations to electronic equipment, furnished as kits whose applicability to specific electronic equipments is identified by type and class. The alterations must be accomplished at the earliest opportunity in accordance with the instructions contained in the kit.

Fire and Bilge Pump. A pump, usually of the reciprocating-steam type, equipped with multiple-suction and discharge manifolds which permit its use for various purposes, such as taking sea suction for fire-fighting, pumping bilges, and ballasting tanks.

Firemain. The salt-water piping system and associated pumps that provide fire-fighting and flushing water.

Firerooms. Compartments containing boilers, boiler accessories, and the station for operating them. When all machinery is located in one compartment, the term used is *machinery room.*

Firesides. The outside area of the boiler tubes which are exposed to the products of furnace combustion.

Firing Aisle. The area between boiler fronts, from which the fuel-oil atomizers are fired and the boiler firing-rate is controlled.

Fleet Training Group. A unit attached to the Fleet Training Command and responsible for working with and grading ships' companies in their performance during refresher training.

Foam. A fire-fighting agent with a high degree of stability and density of structure. It provides a smothering blanket of air-filled foam bubbles.

Foaming. The formation of stabilized bubbles at the water level in the steam drum caused by excessive dissolved solids.

Forced-Draft Blower. A device used to furnish large amounts of air for combustion. Essentially, a large fan driven by a steam turbine or electric motor.

Fresh Water. Both feedwater and potable water.

Freshwater Production. The output of an evaporator which supplies the needs of the galley, washrooms, laundry, boilers, etc. through a series of piping and manifold systems.

Fuel Atomizer. That portion of a fuel-oil burner which breaks the oil into fine particles and injects them into the furnace in a conical spray.

Full-Power Trials. Trials designed to run propulsion plant at maximum rpm, as established by type commanders' regulations.

Galley Steam. Steam provided to steam tables, steam kettles, warming trays, etc.

Gas Turbine. A prime mover consisting of a compressor, combustion chamber, and turbine.

General Quarters. Signal for all crew members to report to their assigned battle stations as quickly as possible.

Gland Packing. Packing designed to prevent leakage of water, steam, or oil along a movable shaft, such as found in steam-turbine rotors, pump rotors, propeller shafts, etc.

Gland Steam. Low-pressure steam introduced into shaft-gland packing to prevent air from leaking into and steam from leaking out of a turbine.

Governor Mechanism. A device used to maintain constant speed of a rotating machine regardless of the load. Sometimes called a speed regulating governor.

Gundecking. Faking or falsifying something, such as a report.

Hardness. The condition of water that is an indication of the amount of scale-forming materials such as calcium and magnesium.

Hardware. The miscellaneous fittings that contribute to the operation of a major unit within a system.

High-Pressure Turbine. A turbine unit that receives steam from the boiler and converts the thermal energy into mechanical energy.

Hotel Services. The collection of ship services that are not related to main propulsion. These services include steam for heating, galley steam tables, and laundry presses; electricity for ship comfort, air conditioning, and washroom facilities.

Hydrostatic Testing. The use of water under pressure to induce a load on a system, thus checking the system above its operating pressure to ensure tightness. Such tests are applied to boilers, condensers, piping systems, heat exchangers, etc.

Impeller. The only moving part of the pump end of a centrifugal pump. It consists of a shaft upon which are mounted radially a series of curved vanes with a suitable means for providing for entry of the fluid. High speed rotation develops a centrifugal force in the liquid.

Inclining Experiment Data Booklet. A two-part book containing in Part I observations and calculations that determine displacement and location of center of gravity of the ship in light condition; in Part II, stability data as related to the characteristics of the ship in operating conditions.

Instructions. Serially numbered directives issued by commanders ashore and afloat.

Integral Superheater. A superheater installed as an integral part of the generating tube bank, rather than in a separate part of the boiler.

Jacking Gear. Mechanism for turning over large turbines without admitting steam to them. It consists of an electric motor, connected through a worm-and-wheel type reduction gear and mechanical clutch, to some part of the reduction gear first-pinion shaft for the main propulsion unit.

Kingsbury Thrust Bearing. An extremely large bearing that absorbs the thrust between the turbine shaft and the propeller shaft.

Leads. Measurements taken by removing upper bearing half, laying several lengths of soft lead wire across the top of the journal perpendicular to the bearing axis, replacing bearing half and setting up tightly, loosening bearing half, and measuring the thickness of the leads at several points with a micrometer to obtain bearing clearance.

Light-Off. The starting of a fire in a boiler. Sometimes applied to the starting of an engine or of the entire propulsion plant.

Lighting-Off and Securing Sheets. Check-off sheets that outline the prescribed lighting-off and securing procedure for auxiliary and main propulsion plants.

Locked Shaft. A shaft that has engaged the jacking gear or a shaft-locking key and cannot, therefore, rotate.

Log Room. The Engineering Department record room on board ship.

Long-Range Combat Effectiveness. The degree of combat readiness for periods of 3 to 5 years in the future.

Low-Pressure Turbine. A turbine, larger than a high-pressure turbine, that receives steam from high-pressure turbine exhaust, removes additional energy, then exhausts the steam to the condenser at a pressure below atmospheric pressure.

Machinery Rooms. Compartments that house both the boilers and propulsion turbines serving a particular shaft.

Machinery Spaces. Compartments that contain the major shipboard machinery under the cognizance of the engineer officer.

Magnetohydrodynamic Process. A process for converting heat to direct-current electricity in which a high-temperature, electrically conductive gas is passed through a magnetic field to induce an electrical voltage.

Main Condenser. A heat exchanger that condenses exhaust steam from the turbines into a liquid state (condensate) at a pressure well below that of the atmosphere.

Main Control. The location at which the EOOW stands his watch and from which he controls the operation of the machinery spaces.

Main Engines. The propulsion turbines.

Main Propulsion Turbine. The propulsion turbine, as distinguished from auxiliary turbines used to drive pumps, forced-draft blowers, etc.

Main Propulsion Unit. The major components that contribute to ship propulsion; propulsion turbine, main reduction gears, and main condenser.

Main Reduction Gears. A speed-reducing apparatus that consists of a set of large gears and pinions.

Main Shaft. The propulsion shaft that runs from the reduction gear aft to the propeller.

Make-Up Feedwater. A distillate that has been in storage in a reserve feed tank, and consequently has greater contamination than distillate and condensate.

Manhole Plates. Plates used to seal the manhole openings.

Manning Level. The percentage of a ship's manpower allowance that is actually on board. In peacetime not figured against complement, but this might be done in wartime.

Manufacturer's Instruction Books. Books furnished by manufacturers of equipment installed.

Maximum Speed. The highest speed obtainable at full-power operation of the main propulsion unit.

Mechanical Energy. Energy that stems from the position or motion of relatively large (molar) masses or machinery. The ratio between it and thermal energy is 778 ft.-lb. to 1 Btu.

Navy Directives. Navy communications in which policy is established or a specific action is ordered; plan issued with a view to placing it into effect when so directed; any communication that initiates or governs action, conduct, or procedure.

Navy Regulations. Principles issued by SecNav and approved by the President for the guidance of the Naval Establishment, particularly the duties, responsibilities and authority of all offices and individuals.

Nuts-and-Bolts Evolutions. Fundamental, daily, work routines.

Officer Of The Deck. An officer on duty, in charge of the ship, representing the commanding officer.

Oil King. The petty officer in the Engineering Department who is responsible for maintaining accurate fuel-oil records, transferring fuel to service tanks, ballasting tanks, lining up the fuel-oil system for internal fuel-oil transfer and for replenishment at sea, and for reporting such data to the engineer officer and the engineering officer of the watch.

Overspeed Trip. A device installed on auxiliary turbines and turbo-generators that automatically shuts off steam to the turbine if the speed-regulating or speed-limiting mechanism (governors) fail and the turbine overspeeds.

Phasing. The process of ensuring that equal loads are maintained on each of the three phases of the electrical system and that phase sequence is maintained to ensure proper rotation of motors.

pH. The degree or acidity or alkalinity of a solution expressed on a scale that ranges from 0 to 14.

Pinch Bar. A lever with a pointed projection at one end, used especially to roll heavy wheels, etc.

Potable Water. Water of drinking quality.

Power Plant Efficiency. Overall efficiency calculated from the ratio between the net cycle useful work and the total heat released by the fuel (based on its high heat value).

Pressure-Fired Boiler. A boiler that operates with a positive air pressure of from 30 psig to 90 psig in its furnace. Normal boilers operate with a slight air pressure (seldom over 10 psig).

Pressure-Fired Boiler With Regeneration. A pressure-fired boiler that heats feedwater by steam extracted expressly for that purpose from various stages of the turbine.

Priming. A mechanical problem caused by slugs of boiler water being associated with the steam coming from the steam drum. Rough weather and/or a sudden increase in steam consumption can bring this about.

Priming Pumps. The process of filling the pump casing and associated piping with the liquid to be pumped.

Propulsion Auxiliaries. The many machines that contribute to the operation of the propulsion system but are not considered part of the main propulsion unit.

Propulsion Unit. The combination of propulsion turbines, main reduction gears, and main condensers in any one propulsion plant.

Rainmaker. Any device used by ship's force to condense shore steam into water in order to save funds when feedwater is needed and the ship's evaporator is inoperative. Use of rainmakers is considered improper by most type commanders.

Range. The distance a ship can travel without replenishment of fuel.

Reduction Gears. A speed-reducing set of gears and/or pinions.

Remote Valves. Valves that can be closed from stations remote from the space where the valve is installed.

Repressurizing. The process of placing a pressure system back in operation after it has been depressurized for repairs.

Restricted Availability. Time assigned for the accomplishment of specific items of work by a Repair Activity, normally with ship present.

Root Valve. A valve located where a branch line comes off the main line.

Safety Valves. Valves installed on boilers to prevent the pressure in the boiler from rising above the safe working limit. They are connected directly to the boiler steam drum and set to release at designated pressures.

Salt Box. A device used by shipyards or large repair facilities to provide a constant, steady electrical load for the testing of ships' service generators.

Saturated Steam. Water heated to its saturation temperature for a specific pressure.

Scale. A hard adhering deposit, caused by excessive sea-salt concentration, which forms directly from the boiler water.

Scoop Injection. A slanted sea chest that permits sea water to flow directly into the circulating waterside of the main condenser, without the need for a pump, as a result of the ship's speed through water.

Sea Chest. Intake in ship's side between hull and first inboard sea valve.

Sea Trials. At-sea operating tests of ship machinery. These may be builders,' acceptance, post-repair, standardization, or tactical trials.

Sea Valve. The first valve inboard of a piping section which penetrates the ship's hull, i.e. the valve just inboard of the sea.

Securing. The orderly, systematic shutting down of the main propulsion plant and/or auxiliary plant.

Settling Out. Allowing fluids to remain stationary in a tank while differences in specific gravity separate sediments and other contaminants.

Shaft Alley. The compartment through which the propeller shafts extend to the propeller, usually just forward of the stern tubes.

Ship Alts. Alterations under the cognizance of the Naval Ship Systems Command involving installed equipment, which change military characteristics of a ship or are made for reasons of safety, efficiency, and economy of operation and maintenance.

Ship's Information Booklet. Five volumes containing system diagrams and manufacturer's data for installed equipment on new construction ships: hull and mechanical; piping, ventilation, heating, and air-conditioning systems; power and lighting systems, electronics systems; and interior communications and fire-control systems.

Ship's Organization Book. An administrative and organizational guide for a particular ship, based on a standard issued by the type commander.

Shore Feedwater. A distillate or condensate from a shore source.

Shore Water. Water from a shore source. It may be potable or not potable.

Short-Term Expediency. The ability to react quickly to a situation, but for a short time only.

Sliding Feet. The bottom flanges of the boiler which are not rigidly attached but are lubricated to permit sliding as a take-up for boiler thermal expansion and contraction.

Sludge. The sediment in the lower portions of a secured boiler resulting from the settling of the boiler water suspended solids. Besides the suspended solids, it may consist of oil and other contaminants. Also, sediment in fuel oil, lubricating oil, etc.

Slugs. Water that condenses in steam lines when the plant is secured and steam lines are not properly drained.

Snipe. A member of the Engineering Department.

Soot Blowers. Stationary and rotating steam pipes installed in boilers to steam-lance soot deposits off the firesides when boilers are being steamed.

Sounding Patrol. Duty men who periodically measure the depth of liquid in the ship's tanks and report results to the engineer officer or the engineering officer of the watch.

Sounding Tapes. Measuring tapes for sounding tanks.

Sounding With Hammer. The process of tapping bolts with a hammer to determine whether they are tightly in place or loose.

Sound-Powered Telephone. A telephone that generates its own power by the sound vibrations of the voice moving in the field of a magnet.

Split Core. A magnetic coupling device used in some electric meters to permit the measurement of current in a wire without having to break any electrical connections.

Sprayer Plate. The part of the fuel-oil atomizer assembly that controls, by its size, the amount of oil entering the furnace and, to some degree, the shape of the oil cone.

Stack Covers. Canvas or plastic covers used to prevent rain and moisture from entering stacks and combining with the sulphur to form sulphuric acid when boilers are secured.

Standdown. A time when a ship is granting maximum leave and liberty and machinery plant is, normally, in cold-iron status.

Steaming Auxiliary. The condition of having main engines secured, but boiler, turbo-generator, required auxiliary condenser, and other auxiliary equipment in operation to provide electrical and other hotel services.

Steam Line. Any piping used to carry steam from the boiler to various pieces of equipment.

Steam Power Plant. A power plant that operates on the basic steam cycle for propulsion, as distinguished from gas turbine, diesel drive, etc.

Stern Tube. A steel tube built into the ship's structure to support and enclose the propulsion shafting where it penetrates the hull.

Stern-Tube Packing Gland. A gland that seals the area between the rotating propulsion shaft and the stern tube.

Superheated Steam. Saturated steam which has been raised to a temperature which is above its saturation temperature.

Superheater. A section of the boiler where saturated steam is turned into superheated steam.

Surface Blowdown. The expelling of a portion of the boiler water from the steam drum.

Suspended Solids. Finely divided and/or large visible particles suspended (not dissolved) in the boiler water or other fluids.

Tag. A card or metal disc secured to a valve or piece of equipment to indicate that it should not be opened or energized without the permission of the person whose name is contained on the tag.

Thermal Energy. Energy characterized by its ability to be transferred from one body to another by temperature difference alone, through the medium of colliding molecules or electromagnetic waves with an attendant exchange of molecular energy.

Thermodynamic Design. A condition wherein all operating equipment functions as an interacting system having a thermodynamic-energy balance.

Thermoelectric Process. A process for converting heat to direct-current electricity utilizing a thermocouple device in which an emf is developed through the application of heat to dissimilar materials.

Thermionic Process. A process for converting heat to direct-current electricity by use of a thermionic converter consisting of two metal electrodes separated by a space under vacuum.

Throttleboard. The display board at the engine control station which includes the main engine throttle valve control.

Throttle Valves. The set of valves used to control the amount of steam entering the main turbine. Also, any valve used to control the rate of flow of a fluid.

Thrust Bearing. A bearing that permits rotation of a shaft while absorbing the transmitted thrust, thus maintaining axial alignment of the rotating shaft.

Top Watchstander. A person who stands the EOOW or EDDO watch or supervises the watch in machinery spaces other than at main control. Also, the man in charge of each machinery-space watch.

Tripping Out. A casualty that causes a generator to secure itself because a sudden overload actuates one of its safety devices, such as the over-speed governor.

Turbo-Generator. Machinery used to generate electric power by means of a turbine prime mover.

Type Command. An administrative subdivision that separates a fleet or force into ships or units of the same type, as distinguished from a tactical subdivision. For example, CruDesPac.

Uniform Code of Military Justice. The code enacted by Congress for all armed services. For the Navy, it replaces the Articles for the Government of the Navy, traditionally known as "Rocks and Shoals."

Uptakes. Large enclosed passages for exhaust gases to transit from the boiler to the stacks. Also called *exhaust trunks.*

Valve. Any device by which the flow of liquid or gas, etc., may be started, stopped, or regulated, by a moveable part that opens or obstructs passage.

Warm-Up. The orderly, systematic procedure for bringing the auxiliary and/or main propulsion plant into full operation.

Watch Officer. An officer regularly assigned to duty in charge of a watch or a portion thereof; for example, the officer of the deck, or the engineering officer of the watch.

Water Hammer. The knocking sound heard in pipes when a large temperature difference exists between the piping material and fluid temperature and causes thermal expansion. Also, the concussion of moving water against the sides of a containing pipe; as in a steam pipe.

Water Jet. A means of propulsion in which water is scooped into an inlet, ducted to a pump located in the hull, and discharged through a nozzle to provide desired thrust.

Watersides. The internal area of boiler tubes, steam drum, header, etc., where feedwater is in contact with the metal.

Watertight Integrity. The condition of having subdivisions, doors, hatches, etc. watertight by maintaining fittings, setting conditions of material readiness, and by training the crew.

Watertight Integrity Test Book. A planned program for conducting watertight integrity tests, including a listing of required test pressures, and places for listing test results and for the signature of person performing the work.

Wear. The wasting away by continual attrition, fluid erosion, or lack of sufficient oil pressure, etc., of a babbitted bearing surface.

Wiping. Failure of a babbitted bearing, in that portions of its surface are completely removed or scored, thus causing the bearing to score the journal shaft or cause the shaft to bind.

RECOMMENDED READING

Official Publications

NavShips *Technical Manual*
NWIP 10-1 *Operational Reports*
NWP 38 *Replenishment at Sea*
NWP 50 *Shipboard Procedures*
NWP 50-1 *Battle Control*
United States Navy Regulations, 1948

Officer Training Manuals and Correspondence Courses:

NavPers 10858C *Engineering Administration*
NavPers 10814B *Engineering Duty Officer*
NavPers 10813A *Engineering Operation & Maintenance*
NavPers 10864B *Shipboard Electrical Systems*
NavPers 16178A *Fundamentals of Diesel Engines*
NavPers 10788B *Principles of Naval Engineering*

There are also training manuals and correspondence courses for enlisted men, and:

Ship Information Book
Damage Control Book
General Specifications for Ships of the U. S. Navy
Inclining Experiment Data Booklet
Military Specifications and Standards

Books

D'Arcangelo, A. M. *Ship Design and Construction.* Society of Naval Architects and Marine Engineers, New York, 1969.

Department of Marine Engineering, U. S. Naval Academy. *Energy Analysis of Naval Machinery.* U. S. Naval Institute, Annapolis, 1940.

———. *Engineering Materials.* U. S. Naval Institute, Annapolis, 1952.

———. *Principles of Basic Mechanism.* U. S. Naval Institute, Annapolis, 1952.

El-Wakil, M. M. *Nuclear Power Plant Engineering.* McGraw Hill Book Company, 1962.

Gill, Paul W. *Gas Turbines and Jet Propulsion.* U. S. Naval Institute, Annapolis, 1952.

———; Smith, J. H. Jr.; and Ziurys, E. J. *Fundamentals of Internal Combustion Engines.* U. S. Naval Institute, Annapolis, 1954.

Johnston, R. M.; Brockett, W. A.; and Bock, A. E. *Elements of Applied Thermodynamics.* U. S. Naval Institute, Annapolis, 1954.

Latham, R. F. *Introduction to Marine Engineering.* U. S. Naval Institute, Annapolis. 1958.

———. *Naval Boilers.* U. S. Naval Institute, Annapolis, 1956.

Payne, C. N. *Descriptive Analysis of Naval Turbine Propulsion Plants.* U. S. Naval Institute, Annapolis, 1950.

Saunders, H. E. *Hydrodynamics of Ship Design,* Vols. I, II, III. Society of Naval Architects and Marine Engineers, New York, 1967.

Seward, H. L. *Marine Engineering,* Vols. I and II. Society of Naval Architects and Marine Engineers, New York, 1968.

Periodicals

ASME Journal. American Society of Mechanical Engineers, New York.

Control Engineering. Reuben H. Donnelley Corp., New York.

Instruments and Control Systems. Instruments Publishing Co., Inc., Pontiac, Ill.

Naval Engineers Journal. American Society of Naval Engineers, Washington, D.C.

Power. McGraw-Hill Inc., New York.

SNAME Transactions. Society of Naval Architects and Marine Engineers, New York.

Welding Engineer. Welding Engineer Publications, Inc., Morton Grove, Illinois.

INDEX

Note: Figures in *italic* indicate illustrations